PIPE LINE CONSTRUCTION

by
Max Hosmanek

edited by
Cinda L. Cyrus

Published by

Petroleum Extension Service
Division of Continuing Education
The University of Texas at Austin
Austin, Texas

in cooperation with

The Pipe Line Contractors Association
2800 RepublicBank Dallas Building
Dallas, Texas

ISBN 0-88698-096-8

Library of Congress Catalog Card Number: 84-60491

CONTENTS

FOREWORD

The officers, directors, and members of the Pipe Line Contractors Association appreciate the opportunity to sponsor this revised publication of *Pipeline Construction — A PETEX Primer.*

A Primer of Pipe Line Construction was first published in 1952 and revised in 1966. Thousands of these publications were distributed throughout the world. The continuing interest and demand for information about this industry is very rewarding. Therefore, as a contribution to both the public and the pipeline industry, the Association, in conjunction with Petroleum Extension Service, The University of Texas at Austin, presents a new expanded version of the pipeline construction industry primer. We hope this presentation will be beneficial in explaining not only how this industry operates but also why transporting materials, liquids, and vapors by pipeline is a reliable, safe, silent, flexible, and cost-effective means of transportation.

Hailey A. Roberts
Managing Director
Pipe Line Contractors Association

PREFACE

Developing from the crude bamboo tubes of the ancient Chinese, modern pipelines cross harsh terrains and link hostile ocean depths to meet the needs of an energy-hungry world. Moving through the lines silently, swiftly, and economically, massive amounts of petroleum products and other materials are transported to consumers around the world. Without the vast and complex pipeline system—and the technology and people who make it possible—the modern world would be a vastly different place.

Despite the vital role that pipelines play in the functioning of the modern world, few people recognize the importance of the pipeline construction industry. Ironically, it is this industry's efficiency, creativity, and technological mastery that mask its invisible but vital role.

With this book, we hope to bridge this gap and provide information about the pipeline industry both to those working in the industry and to anyone interested in pipeline construction. The importance and significance of pipelines is too great to ignore, and by tracing the history of the industry and illustrating the sheer physical effort involved in pipeline construction, we hope to bring about a greater awareness and appreciation of this essential industry.

INTRODUCTION

The story of pipelines and their builders is one of the more dramatic and colorful chapters in the history of modern industrial development. While most people are aware of the rapidity of technological change in today's world, few are aware of the often-invisible underlying structures—such as pipelines—that made this progress possible. The unrecognized role of pipelines in this process is, in a way, a testimony to the effectiveness of their design and function. Unlike other, more visible forms of transportation, pipelines operate silently and unobtrusively, and like the body's circulatory system, pipelines are an unseen but vital supply network and an indispensable tool in the ongoing technological revolution. Just as petroleum and natural gas will be the lifeblood of industry to the end of the century, so too will pipelines continue to function as the veins and arteries providing the crucial link between producers and the ultimate consumers of energy.

Pipelines are now the second largest carriers of intercity freight in the United States. Each day, a vast network of lines moves some 30 million barrels of liquids and millions of cubic feet of gas from coast to coast at rates substantially lower than those charged by other carriers. Continuous expansion has resulted in a network of over 700,000 miles of pipelines.

Comparisons show that pipeline rates are roughly one-fifth of rail rates and about one-twentieth of truck rates. Operating without the benefit of subsidies, pipeline companies have consistently managed to hold down real costs. These lower costs are due in part to the expansion of overall system capacity through the use of highly efficient large-diameter pipe. Costs have also been curbed by utilizing state-of-the-art, automated equipment that requires fewer employees and far less maintenance. Another, more subtle economic factor results from a characteristic unique to pipelines. While other transportation systems require that a product be moved by the carrier, pipelines are specialized fixed structures that utilize pressure or compression to move goods from point to point. Since they serve both as a conveyor and a temporary container for those goods, much of the wear and tear that plague other systems is eliminated, and the savings can be passed on to customers in the form of lower rates.

2

Besides economical transportation, pipelines offer other advantages as well. Because the majority of lines are below ground, pipelines are the most nondisruptive means of land transportation, and no public highways are needed for the movement of pipeline products (fig. 1.1). Also, since they are virtually immune to weather conditions, pipelines can practically guarantee uninterrupted service. Another important advantage of this form of transportation is safety. Pipelines have compiled a safety record that no other system of comparable capacity can even approach (fig. 1.2). This record has been maintained even in the face of increased demands placed on existing systems and a vigorous rate of growth in new construction.

Figure 1.1. This mainline valve is the only visible sign of a vital transportation system.

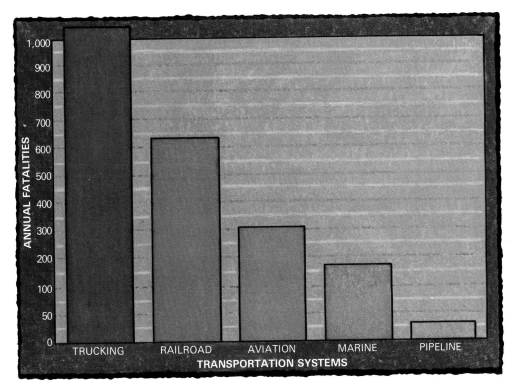

Figure 1.2. As demonstrated by figures for the average annual fatalities for different forms of transportation, pipelines have an impressive safety record.

Projected Growth

Long-range forecasts predict a continuing need for pipeline construction for the rest of the century. Much of the additional mileage will be for the so-called big-inch projects. These projects use relatively thin-walled, high-tensile-strength steel pipe with diameters of 20 inches or more. The great strides made in metallurgy, the science and technology of metals, since 1940 have given the pipeline industry the steels capable of withstanding operating pressures of 1,000 pounds per square inch (psi) or more. With stronger steel, the throughput capacity of the entire pipeline system has actually increased since higher pressure tolerances mean that more materials can be transported.

Pipeline Costs

While operating costs have not increased significantly in recent years, the cost of laying a big-inch line has risen sharply. Construction costs increase approximately linearly as diameter increases, and this is especially true for offshore lines. Consequently, the success of many future big-inch projects will rest on the availability of the massive financing required to carry the project through to completion. With construction costs often running well over $1 million per mile, adequate and timely financing has become as important a variable as weather or terrain. Indeed, the magnitude of much of the planned construction is so great that timing will be decided as much by bankers as by engineers.

In recent years, a number of misconceptions about the impact of pipeline construction and operations on the environment have arisen. The notion that pipelines are environmentally destructive usually stems from the reports of those who witness only the initial phases of construction. The clearing, grading, and ditching that are the first steps to putting the pipe in the ground do necessarily disturb the environment. This disruption, however, is only temporary; the contractor's cleanup crew restores the right-of-way to its previous state and often actually improves its condition (fig. 1.3). Advanced conservation techniques are carefully applied to preserve the topsoil for later replacement. Generally, the long-term effects on the environment are negligible, and pipelines leave little or no indication of their presence. It is not uncommon for cultivation to resume within months of excavation through farmland. This regard for environmental conservation has helped to foster a high level of cooperation between landowners and contractors.

Measures are also taken to ensure that pipe is sound and impermeable to leaks both before and after it is buried. Coatings have been developed to protect the pipe against external corrosion that may cause pipe failure. At the same time, sophisticated corrosion-control devices are placed along the right-of-way to inhibit the electrochemical action that promotes metal corrosion. Together, these techniques are a virtual guarantee against damaging leaks. The entire system is also monitored from an elaborate control center that can quickly identify and isolate a section of the line should a problem develop.

Figure 1.3. Pipelines are safe and silent neighbors.

Products Transported by Pipelines

Since its birth over a century ago, the pipeline industry has undergone profound changes in nearly all of its construction and operation techniques. The number of products carried by pipelines has also grown. Although natural gas, petroleum, and its derivatives still constitute the bulk of traffic, less-traditional commodities such as coal slurry and carbon dioxide are becoming increasingly important.

CRUDE OIL

The pipeline industry originally developed in response to the need for more efficient methods for bringing oil out of the then-booming Pennsylvania fields in the late 1800s. Although crude oil shipments no longer dominate pipeline traffic, they still account for 27 percent, or roughly 190,000 miles, of the total pipeline mileage in the United States. This includes both cross-country trunk lines as well as the shorter field-gathering lines that run to tank storage or to refineries. Energy conservation efforts worldwide during the past decade may result in a continued long-term decline in both the total volume of crude oil shipped and the pace of crude-line construction. Future projects will probably include building more systems in remote areas such as the Arctic and the North Sea. Additional construction may also be needed to transport synthetic fuel products from plants in the Rocky Mountains of the United States (fig. 1.4).

Figure 1.4. Even in the harsh Alaskan climate, pipelines are a vital component of energy transportation.

PETROLEUM PRODUCTS

Products lines are designed to transport refined petroleum derivatives such as kerosine, jet fuel, gasoline, and diesel fuel. Computer-assisted batching allows a number of different products to be moved in the same line, and maintaining a precalculated flow rate prevents contamination between batches. To further guard against confusion and inadvertent misrouting, spheres filled with glycol or water are introduced into the line between batches. These spheres form a seal with the pipe wall and separate the different products in the line. In this way, products may be separated according to type, quantity, and destination. There are currently over 97,000 miles of products lines in the United States, representing approximately 14 percent of the total cross-country pipeline mileage.

NATURAL GAS

Only recently has the full potential of natural gas been recognized. Once flared off at the wellhead as a waste product, natural gas is now being touted as an ideal fuel for today and for the future. In fact, 59 percent of existing cross-country lines, some 410,000 miles, carry natural gas (fig. 1.5). Recent reports suggest that even if world demand grows by 70 percent, gas supplies for the next 20 years will be more than adequate. At current consumption rates, it is estimated that only one-third of world gas resources will be exhausted by the year 2020.

Today the laying of natural gas transmission lines is the fastest-growing sector of pipeline construction. Of the projected cross-country transmission and gathering lines planned over the next 20 years, fully 50 percent are expected to be gas lines. Deregulation and new exploration, in conjunction with a continuing demand, will probably keep gas lines in the forefront of new construction.

Figure 1.5. Natural gas is transported from the well by pipeline.

HISTORY OF THE PIPELINE CONSTRUCTION INDUSTRY

2

Historical evidence indicates that as early as 900 B.C., the Chinese were using bamboo tubes to construct makeshift pipeline systems to carry natural gas (fig. 2.1). The first use of natural gas in the United States, however, did not occur until 1821. Rudimentary wooden pipelines, often no more than hollowed-out logs, carried the gas from what was called the burning

Figure 2.1. Early bamboo pipelines probably resembled these used by the Chinese in Tseliutsing, Szechwan Province.

12

spring to nearby buildings where it was used for light (fig. 2.2). The safety and reliability of these pipelines was precarious at best. It was not until 1843 and the invention of iron pipe that the risk of piping gas even short distances was reduced.

In 1859 the first oilwell was brought in at Titusville, Pennsylvania. This event sparked an increase in drilling and refining activity and spawned a host of petroleum-related industries as well. Transportation, which was needed to provide a reliable system for handling the large quantities of oil

Figure 2.2. An early wooden pipeline fashioned from barrel staves and wire. Leakage and accidental ignition were common hazards.

produced, was one of these. The proximity of rivers and streams to the producing well led to the development of a waterborne transportation system. Using horse-drawn carts, teamsters hauled barrels of crude oil to barge-loading areas. There, other teamsters began the often precarious journey to refineries downstream (fig. 2.3). In addition to the inherent danger and high cost of this method, shippers also risked disruption due to weather conditions and labor disputes. Unfortunately, it was the only method available to them at the time.

Figure 2.3. Almost every conceivable type of craft was pressed into service to carry oil downstream.

Some measure of relief was afforded as the railroads became more heavily involved in hauling oil (fig. 2.4), although railheads were usually located some distance from the wells. Producers were at the mercy of the railroads and the teamsters. The railroads' refusal to allow pipelines to cross their right-of-ways further reinforced this dominance (fig. 2.5).

Efforts to break the hold of the railroads and teamsters proved futile until a group of producers formed the Tidewater System and undertook the construction of the first long-distance pipeline. The Tidewater Pipeline, 108 miles long and 6 inches in diameter, began pumping crude over the Allegheny Mountains to Williamsport, Pennsylvania, in 1879. In Williamsport it was loaded onto tank

Figure 2.4. Early tank cars

cars bound for the New York market. Besides helping to break the railroads' stranglehold, completion of the line represented several engineering milestones. In addition to piping crude a record distance, the line had been built over rugged, mountainous terrain in the dead of winter. Given the construction equipment and engineering techniques at the time, this in itself was a remarkable achievement. What was perhaps most important, however, was that the Tidewater line

Figure 2.5. When faced with the railroads' refusal to grant a right-of-way, this early pipeliner proved to be as resourceful as his modern counterparts. The oil was transferred from the pipeline to carts that carried it across the tracks to holding tanks on the other side. From there it was pumped back into the pipeline and on to the refinery.

16

offered definitive proof of the technical feasibility of long-distance oil pipelines. Twelve years later, the first high-pressure, long-distance gas line began transporting gas 120 miles from the Indiana fields to Chicago.

Pipeline Activity Moves West

Until the turn of the century, pipeline activity was centered around the major production and refining areas in the eastern United States. After 1900 the situation changed rapidly as new discoveries in the West and Southwest were made. The Spindletop gusher in Texas symbolized the potential of the vast fields west of the Mississippi. Subsequent discoveries in neighboring states and in California boosted the population, the production, and the number of pipelines in the region. Refineries in the East increasingly came to rely on these fields, and by 1940 they received 85 percent of their crude from sources west of the Mississippi River.

The economic viability of these and later discoveries depended on whether low-cost and reliable transportation systems could be established between the producing areas and the major markets and population centers in the East. During the 1930s, economic turbulence and intense competition within the industry forced pipeline companies to seek new ways to achieve lower shipping costs. One solution was to increase pipe diameter in order to boost system capacity. Large-diameter pipe was more expensive to lay, however, and required continuous high throughput to ensure maximum operating efficiency. To guarantee the availability of sufficient quantities of crude, pipeline companies established joint ownership of lines. Joint ownership made it possible for shared markets to be serviced with a single line, thereby eliminating costly duplicate lines. Such joint ventures marked a significant turning point in the pipeline industry.

Wartime Pipelines

The outbreak of World War II found the United States sorely lacking the capacity necessary to meet the surge in demand of petroleum created by the war mobilization. This situation was aggravated by the fifty petroleum tankers lost to marauding German submarines in the Atlantic and the diversion of a number of oil tankers to the British early in the war. As a result, average tanker deliveries to the East Coast fell by almost 80 percent. In response to this critical situation, the pipeline industry launched a massive program of emergency construction. In spite of the difficulty in obtaining steel and the shortage of skilled construction personnel, private industry had financed and built over 8,000 miles of new pipelines by 1945.

Working in close cooperation with government planners, leaders in the pipeline industry also helped coordinate a comprehensive supply strategy. Pipeline contractors and oil companies agreed to tear up nearly 1,400 miles of existing lines for relocation elsewhere. Often this involved moving the pipe up to 600 miles. In addition, to accommodate the rerouting of shipments to areas of greatest need, the direction of flow on a number of lines was reversed. Since flow direction and shipment patterns of the nation's pipeline system were already well established, this was a remarkable feat. Crude trunk lines from the fields in Texas, Louisiana, and the mid-Continent ran south to the Gulf and north to refineries in the Midwest. With only 5 percent of oil reaching the East Coast by pipeline or means other than tanker before the war and with the virtual blockade imposed by submarine activity, a massive realignment of the system was critical.

In the United States, the most notable wartime project was the government-financed construction of the War Emergency Pipelines (WEP). WEP was a nonprofit corporation composed of eleven oil and pipeline companies organized under government sponsorship. Beginning in 1942, WEP set out to construct the world's largest-capacity, cross-country petroleum pipelines. The first of these, dubbed Big Inch, was a crude oil line with a 24-inch diameter (fig. 2.6). Big Inch engineers were confronted with unique design problems in laying the 1,340-mile line. For example, the engineers were forced to rely on relatively untested seamless pipe to construct the line because most steel plate had been requisitioned for warships. More than 16,000 men and 100 contractors eventually began work on the line. Moving at a record rate of 5 miles per day, the line was completed in just over a year, and soon nearly 300,000 barrels of oil a day were flowing to New York and Philadelphia.

Figure 2.6. As unit trains carrying oil sped east, pipeliners raced to finish the Big Inch crude oil line.

18

During the same period, a 20-inch products line, known as Little Big Inch, was laid from Texas to New Jersey in just 370 days, and it transported 250,000 barrels of products a day. When fully operational, both lines accounted for 75 percent of the petroleum and products being delivered to the East Coast in 1944 (fig. 2.7).

Pipelines were also indispensable to the overseas war effort. American and European engineers and technicians used their expertise to overcome seemingly insurmountable obstacles. For example, the famous Flying Tigers in Burma were supplied by a 3,000-mile pipeline that ran from Calcutta to Southwest China, crossing some of the most formidable mountain terrain on earth. Completion of this line was a lasting monument to the skill and ingenuity of the members of the pipeline industry.

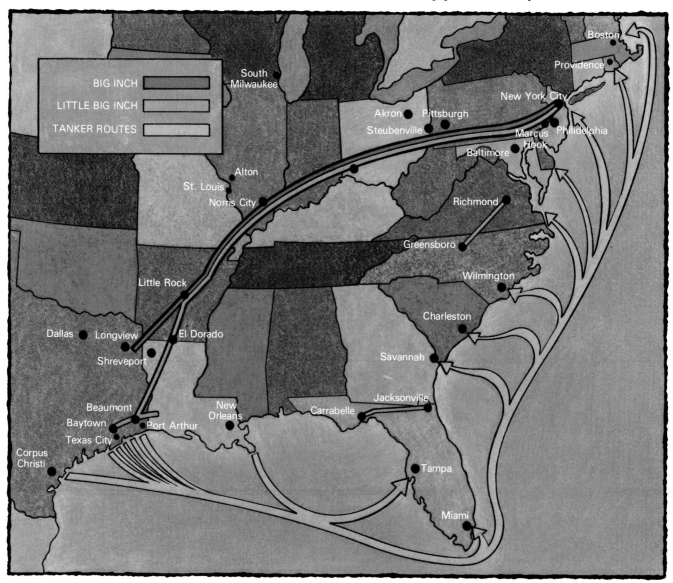

Figure 2.7. Big Inch and Little Big Inch formed the backbone of an integrated transportation network designed to move crude oil and petroleum products to the East Coast. Oil from Texas wells—traveling at 4.5 mph, 108 miles per day—reached Philadelphia refineries in 12 days.

In the European theater, equally difficult conditions were overcome in supplying the Allied effort. To supply troops on the Continent, British engineers modified the techniques originally designed for subsea cable installation to lay pipe. Under the code name "Operation PLUTO" (Pipe Lines under the Ocean) specially reinforced small-diameter pipe was laid out from a cable-laying vessel in 30-foot lengths from huge reels, while both the Germans and the notoriously fickle channel weather were battled. Twenty of these lines, with such unlikely code names as Bambi and Dumbo, were eventually placed under the English Channel to France.

After the war, the WEP lines were sold to private companies, which converted them for gas transmission use. The experience in building these lines, however, had revolutionized the economics of pipeline construction. These wartime projects proved conclusively that the economy-of-scale principle was as valid for pipeline transportation as it was for other industries. They had also proved that it was feasible to move large volumes of refined products and crude oil from producing and refining centers to multiple market areas. Although the advantages of larger diameters had been recognized before World War II, traffic in most cases was insufficient to justify the high construction costs. When additional capacity was desired, the usual procedure was to lay an additional line alongside the existing one. This technique, called looping, increased maintenance costs considerably. By contrast, big-inch lines demanded substantially greater initial outlays of investment capital. This investment, however, was offset by reduced operational costs in the long run.

The experience gained during the war years provided the impetus for the industry's rapid postwar expansion. In the ensuing decades, pipeliners met the challenge of an increasingly industrial nation by continually surpassing their own records of length and size of lines stretching across the continent. Showcasing this growth is the free world's largest pipeline, the Trans-Alaska Pipeline System, a 48-inch line stretching approximately 800 miles.

TECHNOLOGICAL INNOVATIONS: A HISTORICAL OVERVIEW

3

The discovery of oil in the United States and the ensuing rapid growth in demand for petroleum products helped lay the foundation for the pipeline construction industry. In the emerging industrial centers of the Midwest and Northeast, pipelines came to be relied on more and more to provide the energy needed to sustain the momentum of this economic expansion. As a result, the pipeline industry—aided by the remarkably quick development and adaptation of a number of engineering techniques and innovations—matured rapidly.

The steps in building a pipeline have remained essentially the same since the beginning of the industry. Design variations to accommodate differences in terrain, climate, and operator specifications are frequently necessary, but these do not appreciably alter the construction sequence, which is basically the same for all types of pipelines. Refinements are also constantly being implemented in an effort to upgrade efficiency and safety.

It requires but a brief look at the range of equipment used by the first pipeliners to appreciate the difficulty of their task and the ingenuity and

determination they showed in successfully accomplishing it. Early pipeline construction was arduous and backbreaking labor, and many of the tools used were primitive by modern standards. It took a combination of human and animal power to push through remote and often hostile environments, and workers endured hardships reminiscent of those experienced by the pioneers (fig. 3.1). Undoubtedly these experiences helped forge the pipeliners' well-deserved reputation for independence and rugged self-reliance.

Figure 3.1. Oxen and horses were often used in the construction of early pipelines.

The majority of changes that have occurred in pipeline construction are mainly the result of increased mechanization. Improvements in machine design and function have yielded significant savings in both time and money without any sacrifice in quality. Operations such as ditching used to require hundreds of laborers working with picks and shovels. Modern wheel ditchers can excavate miles of right-of-way in a fraction of the time previously required (fig. 3.2).

Even the early pipeliners recognized that the line must conform to the general contours of the right-of-way. To contour the line, they had to bend joints just as they do today. One of the earliest methods was the fire bend. The first step in making this bend called for the joint to be placed over a small bonfire (fig. 3.3). When the heat had rendered it sufficiently malleable, the joint was placed against a tree, and pressure was applied until the desired bend was achieved. These fire bends significantly weakened the pipe and increased the likelihood of pipe failure. Another early method for bending the pipe called for bracing the pipe so that one end was elevated. Weight was then added to the raised end—usually by having workers stand on the pipe—until the bend was achieved (fig. 3.4). This method was haphazard at best and, like fire bends, weakened the pipe.

The modern bending process, termed *cold working* to contrast with the fire-bend process, is also mechanized. Track-mounted, hydraulic pipe-bending machines literally stretch and thin the pipe wall at the bending point. Paradoxically,

Figure 3.2. Forerunner of the modern wheel ditcher

Figure 3.3. Preparing for a fire bend

this stretching actually increases the yield strength of the pipe, and highly accurate bends can be made.

The evolution of the pipe-joining process is perhaps one of the most dramatic examples of the magnitude of the changes that have transformed pipelining. The crucial operation in pipeline work is the joining of the pipe ends. Before the introduction of oxyacetylene welding, crews used large wrenchlike hand tongs to connect threaded pipe ends (fig. 3.5). Then the entire length of pipe had to be rotated

Figure 3.4. Adding weight to one end of an elevated joint to achieve a bend

Figure 3.5. Early 20th century tong gang

to make the connection. In 1912 a significant improvement in this operation came about with the introduction of the pipe-screwing machine. Its obvious advantage was that pipe ends could be joined more quickly and with less labor. However, a major problem still remained: both manual and mechanically joined pipe joints were extremely unreliable and leaked under normal operating pressures. Until technology could solve this problem, pipeline operators were forced to maintain large crews whose sole purpose was to locate and repair such leaks.

Steel pipe became commonplace in pipeline work in the early 1920s when advances in metallurgy made it possible to produce a grade of steel capable of maintaining its strength and ductility when subjected to the extreme heat generated by the welding process. The development of this type of steel paved the way for the use of oxyacetylene welding. This technique revolutionized pipeline construction by providing a more secure solid bond than any previously available. The high temperatures necessary to complete the process literally melted the metal of the adjoining pipe ends, fusing them together. It combined a relatively simple technique with increased safety and reliability in the finished work. Although oxyacetylene welding represented a significant improvement over the traditional mechanical pipe-joining methods, it too had disadvantages. Difficulty in controlling the depth of the weld resulted in the seepage of molten weld material onto the interior surface of the pipe. This led to corrosion problems and eventually resulted in higher operating costs.

An answer to the problem was found with the introduction in 1928 of electric-arc welding (fig. 3.6). The first field use of electric-arc welding roughly coincided with the initial appearance of 40-foot lengths of seamless pipe. The simultaneous introduction of these two innovations proved to be mutually advantageous. Control of weld depth, now more precise than before, helped minimize internal friction caused by overpenetration. In addition, the greater tensile strength afforded by the seamless pipe greatly enhanced both the safety and the efficiency of pipeline operations and foreshadowed the later development of large-diameter, high-pressure pipelines.

Subsequent advances in welding techniques have essentially focused on modifying the electric-arc process. For example, inert gases are now used to shield the arc from the atmosphere, thereby reducing oxidation and producing a tougher, more ductile weld. Automatic welding machines are also used in situations requiring a steady welding rate and a high degree of consistency.

Similar progress has occurred in every other phase of pipeline construction. Although much of this progress has been gradual and specific changes have often been quite subtle, the cumulative effect is nonetheless impressive. The great strides that have been made in the application of pipe coatings, for example, have significantly reduced the probability of pipe failure due to external corrosion. Instances of damage to external coatings due to careless handling have also been reduced, thanks to more secure methods of lowering-in.

Figure 3.6. Electric-arc welding in the 1930s

Differences are also readily apparent even in the less glamorous — but no less important — aspects of pipeline construction. Communications have been infinitely improved by the simple addition of two-way radios. This minor innovation has made it possible to coordinate the activities of hundreds of workers spread out along miles of right-of-way. Behind the scenes, computers are being integrated into virtually every aspect of the job. Programs have been developed for pipe stress analysis, hydraulics calculations, deepwater pipe laying, and various other design and construction problems. Space research also has relevance for pipeline work; some companies are experimenting with satellites to determine optimal route selection and to aid in mapping routes.

The pervasive effects of technology have added to the complexity of laying a pipeline, but as in all modern industries, adaptation is necessary for survival. Yet despite this trend toward mechanization, the pipeline construction industry still retains an element of adventure that has been lost in many other enterprises. It was in this spirit that pipelining was begun over a century ago, and it is this spirit that makes pipeline construction the art and science that it is today.

MODERN PIPELINE CONSTRUCTION

4

Support Services and General Crew

The *spread*—that is, the necessary equipment and crew needed to build a pipeline—is in essence a moving assembly line (fig. 4.1). Modern spreads, which can consist of 100 pieces of equipment and over 500 workers, are a far cry from the horses and gangs used in the early days of pipeline construction. As is true of most large-scale endeavors, modern pipeline construction requires extensive logistical and administrative support if it is to function at maximum efficiency. Competent and dedicated professionals working behind the scenes coordinate and administer the internal operations of the spread to keep it running smoothly. Without their efforts, no progress would be possible.

Running a spread is the responsibility of the spread superintendent. This individual must have a wide range of abilities and excellent managerial skills. As the overall authority in the field, he represents the contractor's interests. He is empowered to make the critical decisions that will determine the outcome and thus the profitability of the project.

The assistant superintendent and the gang foremen supervising the various crews make up the field administration of the spread. Like troop commanders, they must rely both on technical expertise and a talent for personnel management in order to be effective. The results of their efforts are apparent in the safe and efficient completion of every aspect of the job.

The responsibility for the contractor's financial affairs on the spread rests with the field office manager. He too must be a multitalented individual capable of overseeing billing arrangements, payroll, and any other money-related contingencies that may arise. By closely monitoring expenditures, an astute office manager can significantly reduce the overall cost of a project and bring it in under budget.

The effectiveness of the administrative side of a pipeline construction project largely hinges on the talents of those who service the machinery that actually performs the construction. Besides mechanics, parts and warehouse personnel are also vital. Even the most highly skilled mechanics are of little help if the correct part or tool is unavailable. Finally, there are the truck drivers and other support personnel who perform less-visible but still indispensable tasks. They keep downtime low and productivity high by regularly delivering essential fuels and lubricants and by conscientiously following regular maintenance schedules. Together, all of these workers form the foundation for effective and efficient pipeline construction.

Figure 4.2. Aerial maps show terrain features along the route.

Figure 4.3. A strip map is composed of data from a number of aerial maps; the map in figure 4.2 is included in this strip.

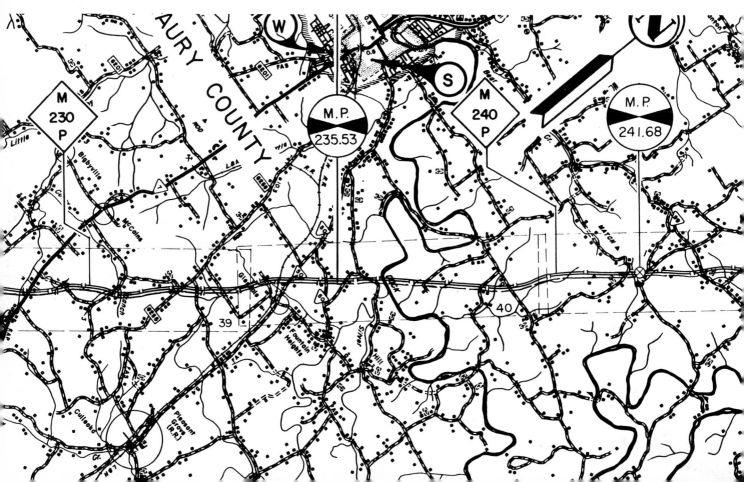

Right-of-Way

ACQUISITION

Construction of a pipeline is a major and expensive undertaking, and meticulous research and planning are needed before a project is started. One of the first things that must be determined is whether or not a market for the products the pipeline will carry exists in the intended service area. Equally important is the necessity of securing an adequate and reliable source of the product to guarantee long-term supply. Possible routes are surveyed by aerial photography and surface mapping (figs. 4.2, 4.3). After careful study, a feasible route must be selected.

The owner company is responsible for handling the legal obligations involved in procuring the proposed route. This may entail acquiring right-of-way, which is the legal right of passage over public land and privately owned property. While land acquisition is an obvious first consideration, multiple local, state, and federal regulations often make the procedure less than straightforward.

The project's planned route can influence acquisition efforts. For example, the closeness of the proposed pipeline to population centers, highways, and power lines may mean that the route must be altered. Even when such route changes are not called for, the pipeline company must still obtain permits from the appropriate authorities in order to cross rivers, streams, roads, and railroad right-of-ways. Another complicating factor is that the proposed route may be under the jurisdiction of several agencies with overlapping authority.

For example, federal regulations for a proposed route may be administered by the Departments of Energy and Transportation as well as the Federal Energy Regulatory Commission, the Interstate Commerce Commission, and the Environmental Protection Agency, to name just a few. Of course, many pipeline right-of-ways now parallel existing pipelines or power lines or other public corridors (fig. 4.2). In some situations, this simplifies the preconstruction phase by eliminating protracted negotiations with a large number of property owners.

The environmental effects of all large-scale construction projects—not just pipelines—have generated much debate and publicity. Largely as a result of this increased awareness, one of the most important prerequisites for construction is the now-mandatory environmental impact study. Pipeline companies have been the leaders in this area, implementing engineering innovations designed to minimize the long-term effects of pipeline construction on the environment. Engineers on the Trans-Alaska Pipeline, for example, devised an ingenious system of vertical support members (VSMs) that effectively prevented the warm oil carried in the aboveground pipeline from melting the permafrost beneath the fragile tundra (fig. 4.4). Similar innovations on other projects serve to underscore the commitment of both contractors and owner companies to meeting energy needs while at the same time safe-guarding the environment. Pipeline contractors are also aware of the possibility that their excavations may uncover potentially valuable archeological sites. In such cases, experts are

Figure 4.4. Vertical support members (VSMs) were designed specifically for the Trans-Alaska Pipeline to prevent thawing of the permafrost.

called in to supervise the preservation of what are often priceless historical and anthropological artifacts.

CONTRACTOR'S COMPLIANCE WITH LANDOWNER'S REQUESTS

Negotiating with the owners of land that the pipeline will cross is a vital step in the right-of-way process. The written agreement, called an *easement,* negotiated by the pipeline owner for use of land along the right-of-way serves as a framework for subsequent agreements worked out between the contractor and the individual property owners. These for-

Figure 4.5. Timber cleared from the right-of-way is neatly stacked for later use by the landowner.

malities precede actual construction, and accommodation of the landowner is the goal of these negotiations. Every effort is made to demonstrate the mutual benefit of the pipeline to both parties and to establish a cordial relationship between the future neighbors. Both the pipeline owner and the contractor recognize the importance of gaining the trust and cooperation of the property owners who will be indirectly instrumental to the successful completion of the pipeline.

A landowner often has specific requests that must be honored by the contractor, and these frequently impose conditions on the use or disposal of material cleared from the path of the right-of-way. For example, an owner might request that timber cut along the right-of-way be sawed into usable lengths and stacked in an accessible location (fig. 4.5). If there are no such stipulations and if local ordinances do not prohibit it, cleared timber and brush are usually pushed into a pile and burned. Similarly, any rock unearthed by clearing or blasting operations must be disposed of, especially in cultivated areas.

Pipeline work necessarily involves the excavation of large amounts of soil. Recognizing the value of this resource, the pipeline contractor must make a concerted effort to implement soil conservation measures. Many contractors now make an initial shallow run with a ditcher in order to remove the topsoil only. This technique, known as topsoiling, places the topsoil in a spoil bank separate from the rest of the excavation (fig. 4.6). Later, during backfilling, the valuable topsoil and other spoil can be replaced in their original strata.

Figure 4.6. Valuable topsoil has been separated from the rest of the spoil bank so that it can later be replaced.

ACCESS TO RIGHT-OF-WAY

While the pipeline owner is charged with securing the land on which to build the pipeline, it is the contractor who must determine how the workers and equipment will gain access to the line. Existing roads that intersect the pipeline route are the usual means. In their absence, however, special access roads known as *shooflies* must be built to link the right-of-way with existing roads. Again, the landowner's wishes must be respected, since the heavy use of access roads can be very disruptive. In addition, laws governing the movement of heavy trucks and equipment on state roads must also be considered. After construction is completed, the landowner may elect to either leave the road or have the land restored to its original state.

FENCING

A pipeline right-of-way spanning open rangelands and fields constantly passes through fenced property. Since fences serve both to mark boundaries and to contain animal herds, these functions must be maintained during the actual construction. Once access has been arranged, the contractor is responsible for controlling movement into and out of the right-of-way. To this end, a fencing crew is sent in to construct temporary gates at points where the right-of-way crosses the fence line. Keeping gates closed when they are not in use ensures both the continuity of the fence line and the goodwill of the landowner.

CLEARING

The width of the right-of-way varies according to contract specifications and individual easements. Generally, it is between 50 and 100 feet, depending on the size of the line and the terrain. While the pipeline itself will occupy only a small portion of that area, adequate space is required on both sides of the ditch for the movement of equipment and personnel and for excavated spoil. As implied by their name, members of the clearing crew transform the right-of-way into a suitable work area by clearing away brush and other vegetation. Unless the land for the proposed route is in an area with no vegetation or encompasses an already cleared right-of-way, or corridor, a clearing crew is needed. Contract specifications, state and local regulations, and the property owner's easement all affect the nature of the disposal of the brush and timber encountered on the right-of-way. Generally, the accepted practice is to cut the trees, save any marketable timber for the property owner, and dispose of the remaining brush and debris by stacking it in piles and burning it in compliance with state and local regulations.

Because of their versatility, bulldozers are used to handle most clearing chores. Chain saws have replaced axes for clearing larger trees, and stumps are either pulled out by bulldozer or removed with explosives. Right-of-ways traversing swamps or wet areas usually need some form of soil stabilization, which is done by riprapping, or spacing logs and timbers evenly along the length of the right-of-way. Installing culverts may also help alleviate some of the problems caused by insufficient drainage. Clearing is completed when all aboveground obstructions judged to be impediments to construction have been removed.

GRADING

The purpose of grading is to provide a smooth and even work area and to facilitate the movement of equipment onto and along the right-of-way. Grading usually entails much more, however, than simply leveling the right-of-way. High- or low-ground areas may require cutting down or filling in to achieve a more uniform grade for the pipeline (fig. 4.7). While rock close to the surface can be ripped out (fig. 4.8), deeper formations and large

Figure 4.7. In rolling terrain, both cuts and fills are required to build the right-of-way.

Figure 4.8. A bulldozer equipped with a ripper attachment dislodges rock.

rocks and boulders often have to be blasted ahead of grading operations (fig. 4.9). As in all situations where explosives are necessary, safety is the prime consideration. Swamps or areas with poor drainage may require the diversion of small streams and the construction of embankments to contain their flow.

Hauling Pipe to the Right-of-Way

The owner company furnishes the pipe and transports it to a central storage and distribution point such as a railhead or warehouse yard (fig. 4.10). Once the pipe is at the distribution center, the contractor must arrange for it to be strung along

Figure 4.9. When explosives are needed in right-of-way grading, drills are used to make holes for the explosive charges.

Figure 4.10. Railroads are the major carriers of pipe to central distribution areas near the right-of-way.

the right-of-way. If the distribution point is reasonably nearby, the contractor may elect to do the stringing himself if suitable equipment is available to haul the individual pipe joints. Because of the specialized nature of pipe hauling, many contractors find it advantageous to subcontract this phase of the project to a carrier better equipped to handle such work. These specialized companies are more familiar both with the requirements of the job and with the state and local regulations governing heavy hauling. When the distribution point is located a considerable distance from the right-of-way, the contractor often hires a trucking firm. In extreme circumstances where the right-of-way is inaccessible to ground transport because of terrain, helicopter stringing may be the only feasible alternative. Such situations, which are rare, are extremely expensive and are contracted out.

Regardless of how the pipe is moved, extreme care must be exercised during loading and unloading. Rough handling, which can gouge or dent the pipe, may result in the need for costly repairs. Coated pipe warrants extraordinary precautions because of its increased susceptibility to deformation. Heavy-duty cranes are usually employed for loading the pipe. Special curved aluminum plates at the end of the wire rope slings are placed inside the pipe ends so that contact with the coating and the beveled ends of the pipe is kept to a minimum. Nylon slings can also be used for carrying the pipe short distances. As the pipe is lifted off the trailer, the weight of the load must be distributed evenly so that the joint does not buckle (fig. 4.11). A spreader bar positioned between the two lifting lines is used for this purpose, especially with large-diameter 80-foot joints.

Figure 4.11. A stringing side boom unloads pipe and places it along the right-of-way.

Trailers for hauling line pipe, such as 80-foot steerable pole trailers, are specifically designed to maneuver the load safely both in traffic and on the temporary access roads into the right-of-way. Once on the right-of-way, hauling trucks may require a tow tractor for extra power on steep grades and on wet or icy surfaces (fig. 4.12). If the right-of-way is close to a well-maintained road, a boom truck can be used for hauling and stringing.

Figure 4.12. When bad weather or steep grades limit access to the right-of-way, heavy equipment is used to assist the stringing operation.

Stringing Pipe

Stringing is the delivery and distribution of line pipe where it is needed on the right-of-way and when it is needed. As with grading and clearing, timing is vital, and the importance of coordinating this activity with other work on the spread is the key to an efficient operation.

Although the grade of steel and the wall thickness of the pipe used on the job is generally uniform, the contractor must be alert to any special requirements and must string accordingly. Joints of special wall thickness and pipe grade are placed at specific locations such as road crossings and wherever else heavy wall thickness may be specified by the contract or by regulations. In locations where there is frequent movement of livestock or vehicles, pipe is strung in a manner that does not obstruct their passage. Since pipe stringing usually precedes ditching, allowances must also be made for the later passage of ditching equipment (fig. 4.13).

Figure 4.13. Pipe joints are strung so that they are readily accessible but do not obstruct the movement of equipment and personnel along the spread.

Ditching

Ditching is more than simply excavating a trench in which to lay the pipe. Since the ditch in which the line is laid affects both the safety and the service life of the pipeline, a number of factors must be considered when the ditch is designed. One of these is the depth at which the pipeline must be buried, known as its *cover requirement*. While this depth is generally standard over most of the line, it usually is greater at railroad and highway crossings, for example, because of the unusual loads and the amount of stress these locations receive. Related factors such as pipe size and soil type also figure prominently in ditch-design calculations.

Ditching standards are formulated by the U.S. Department of Transportation based on soil type and pipeline location. This code prescribes minimum cover requirements for all types of pipeline construction. Where local ordinances differ from these guidelines, an exception is made, and the local authority prevails. Standards specifying ditch dimensions and cover requirements are legally binding on a contractor.

DITCHING EQUIPMENT

No one piece of ditching equipment or particular method will perform equally well under all conditions. Wide variations in terrain, geology, and weather are commonly encountered from job to job and often even along the same stretch of right-of-way (fig. 4.14). These variations mean that the task of determining the machinery and the methods best suited for the particular job must be met with flexibility and innovation.

Figure 4.14. Variations in topography affect both the rate of ditching and the type of equipment required.

For uncomplicated ditching operations in stable soil, the workhorse of the industry is the wheel ditcher. The *wheel* in this case refers to the large, rapidly rotating set of toothed buckets that lift dirt out of the ditch and feed it into a conveyor mounted on the side of the machine. The excavated dirt, or spoil, is then neatly piled to the side of the ditch to facilitate rapid backfilling after the pipe is laid (fig. 4.15). The wheel can be adjusted so that the buckets maintain a constant digging depth, and experienced operators can often help reduce the amount of pipe bending needed by flattening out irregular areas on the ditch bottom. The angle of the ditch walls is purposely sloped outward to decrease the danger of a cave-in while workers are in the ditch.

It is not unusual for different pipeline routes to intersect or for the ditching crew to come across other buried structures in the path of or along the ditch line. Since the contractor is liable for any damage he inadvertently causes, he must use extreme caution in order to avoid disturbing other buried equipment. Besides pipelines, utility cables and other special-purpose subterranean construction such as drain tile are also frequently encountered (fig. 4.16). Where the right-of-way passes through such areas, different equipment such as backhoes, which are relatively more precise, may be needed to excavate a particular section.

The many different types of rock and their varying degree of hardness necessarily makes the term *rock ditching* a catchall that covers a wide range of possibilities. When working with some of the relatively lighter, softer rocks such as limestone or caliche, it is

Figure 4.15. Modern wheel ditchers leave a neat spoil bank that aids in rapid backfilling.

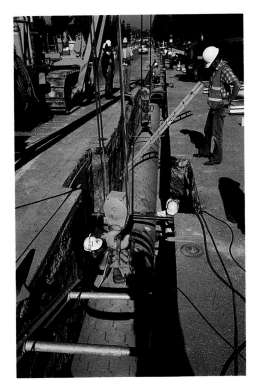

Figure 4.16. Ditching in urban areas is considerably more complicated than it is in rural areas.

40

Figure 4.17. Large pneumatic drills bore through rock to make holes for the explosive charges.

possible to modify standard ditching equipment so that it will break through more easily. Harder rock and more densely packed formations usually require specialized techniques and equipment.

Blasting, or shooting rock, has long been the standard remedy for dealing with rock in the ditch line. The use of explosives is not the imprecise, scatter-gun operation it may at first seem to be. Highly skilled experts determine the optimal placement of the charges in order to maximize the effects of the blast. Large air compressors power the rock, or wagon, drills used to bore the charge holes (fig. 4.17). If there is a possibility that loosened rock may be scattered over the right-of-way or adjacent property, blasting mats are laid out in order to hold down flying debris. The use of explosives does, of course, present certain inherent risks, but in experienced hands, explosives are an invaluable, labor-saving tool (fig. 4.18).

Figure 4.18. Although the effects are often spectacular, blasting operations are strictly controlled in order to eliminate unnecessary cleanup.

Recent advances in metallurgy have produced alloys of such toughness that it is now possible to run ditchers through certain types of rock and rocky soils where explosives would have been required before. Conical teeth made of superhardened carbide or tungsten steel can be substituted for standard ditcher teeth for heavy rock work. Expanding on technology first developed for use in the oil industry, engineers have also succeeded in devising a way to maintain the requisite amount of pressure, or weight, on the digging wheel. This enables the special teeth to penetrate the rock to a depth sufficient for breaking and excavating (fig. 4.19).

Figure 4.19. Designed to dig through rock and rocky soil, this wheel ditcher incorporates the latest technological innovations.

Other pieces of equipment are available for ditching through rock. The most versatile of these is the backhoe. Not only can it clear blasted rock out of the ditch, but it is an excellent excavating machine in its own right. Where large rocks are loosely packed within the soil, a backhoe can usually loosen and remove all but the largest boulders (fig. 4.20). Bulldozers with ripper attachments can also be run along the ditch line before excavation. The ripper's claw-shaped plow helps loosen rock and locate solid formations that may require shooting (fig. 4.21).

Figure 4.20. In the hands of a skilled operator, backhoes have a remarkable range of applications.

Figure 4.21. In rocky soil, a ripper aids ditching excavation.

DITCHING IN STEEP OR MOUNTAINOUS TERRAIN

Ditching on inclines or steep grades requires no special equipment—just more of it. A common technique is to use cables to tie the working backhoe or ditcher to several tow cats positioned at different levels along the right-of-way above it. Tension on the cables enables the equipment below to maintain the position necessary to dig efficiently. These tractors gradually tow the equipment up the grade as ditching progresses.

DITCHING IN WATERLOGGED OR UNSTABLE SOIL

Problems encountered while ditching in rocky soils or solid rock stem from penetration difficulties. Other types of soil, however, present just the opposite problem, namely extreme softness or fluidity. Sandy soils and those containing quicksand, as well as soils that are normally stable but become muddy quagmires when wet, are included in this category. (Procedures for ditching in actual swamp lands and marshes are detailed in Chapter 5, Specialty Construction.) In most instances, soil instability results from too much water, regardless of whether this was caused by a sudden downpour or is a normal characteristic.

When faced with working in waterlogged soils, the contractor may opt to install well points to dewater the ditch line. Well points can effectively dry out extensive areas along the ditch line by continuously removing underground water. Well points function in essentially the same manner as submersible pumps. Every few feet along the ditch line, hollow steel rods approximately 18 feet long are driven into the ground to a specified depth. The buried end is covered with a fine wire mesh that screens out dirt and debris. On the top end, a flexible hose or tubing connects the well point to a header pipe that carries the collected water back to the portable well-point pumps (fig. 4.22). These powerful units maintain suction in the well points and also pump the water away from the ditch. Disposal of the water far enough away from the job site is important since the water may return once it drains back into the underground water table.

Hundreds of well points can effectively stabilize soil that could not otherwise stand on its own. Thus, they may offer significant savings in both time and money in wet areas. They are particularly useful in areas where conventional pumping methods have been ineffective in establishing lasting stability of the ditch walls.

Figure 4.22. Subsurface water is pumped off by a well-point system in order to stabilize the ditch.

DITCH-WALL STABILIZATION

Even with well points, effective drainage is sometimes simply not possible. Where there is a danger that ditch walls may collapse, sheet piling or other materials may have to be used for shoring (fig. 4.23). Besides safeguarding personnel working in the ditch, shoring also prevents undermining the foundations on adjacent structures or roadbeds.

DITCHING FOR CREEK AND STREAM CROSSINGS

Small creeks and streams pose slightly different problems for ditching operations. Cover requirements are normally increased to 5 feet, necessitating the removal of significantly more spoil. Large flume pipes are usually installed to allow the creek to flow normally and to provide a pas-

Figure 4.23. Sheet piling is used to shore the ditch.

sage for equipment over the water. Backhoes or draglines are then used to excavate the ditch across these small streams (fig. 4.24). The specific procedure for excavation and installation depends on the particular permit requirements issued by the governing body for that particular waterway.

Figure 4.24. Backhoes or draglines are used to excavate the ditch in small streams.

Pipe Bending

If the majority of pipeline routes crossed only flat areas or plains, pipeline construction would be considerably less complex. Such is rarely the case, however, and pipeline contractors must contend with a variety of topographical features ranging from gently rolling hills to entire mountain ranges. The permanence of all but the most minor of these obstacles forces the contractor to adapt the pipeline to the terrain through which it passes. Only rarely is pipe ever lowered into the ditch without some deliberate alteration of its longitudinal profile. Although the ditch bottom may be relatively flat, it still reflects the general contours of the right-of-way, and the pipe must be bent accordingly.

In contrast to the fire bends formerly used to bend pipe, the process now used is termed *cold working*. As the name implies, this method acts in the absence of heat to literally stretch and thin the pipe wall at the bending point. Paradoxically, this stretching actually increases the yield strength of the pipe. High yield-strength pipe often resists bending, and it may spring back after each pull is made. In such cases, the machine operator must be especially careful to make the bend as smooth as possible to avoid damage to precoated pipe.

A skilled operator using a track-mounted, hydraulic pipe-bending machine can now tailor the shape of the pipe to conform remarkably well to the contours of the terrain (fig. 4.25). The actual bend is made by a set of clamps, or shoes, that grip the outside surface of the pipe at the point where the bend is to be made (fig. 4.26). These clamps also immobilize the pipe to prevent slippage. A winch cable hooked to the free end of the pipe maintains upward pull during the bend and guides the pipe through the machine. Instructions written on the outside of the pipe by the bending engineer indicate the type of bend and the angle required. To prevent distortion of the pipe, an internal mandrel is used when necessary to maintain roundness in the bending area.

Figure 4.25. Modern pipe-bending machines shape pipe joints to lie evenly in the ditch.

Figure 4.26. Jaws of the pipe-bending machine

Most contracts stipulate a minimum distance allowed between bends and welds and from the end of the pipe. In addition, the maximum allowable degree of the bend usually depends on the diameter of the pipe. For example, a contract may stipulate that the maximum angle of a bend cannot exceed 1.5 degrees in a length of pipe equal to its diameter.

Road Crossings

Contractors strive to keep pipeline construction as unobtrusive as possible while still maintaining the industry's high standards of safety and productivity. When pipeline work presents the potential for disruption, such as when a new line must cross a road, attempts are made to minimize any inconvenience by proceeding as rapidly as possible with the work. Thanks to major engineering improvements over the years, road crossings now represent one of the best examples of the contractor's regard for the safety and convenience of the public. Of the techniques commonly used, most involve some sort of horizontal boring beneath the roadway (fig. 4.27). Thus, laying a pipeline under a heavily traveled thoroughfare now involves virtually no interference at all with traffic flow nor any damage to the road or roadbed.

Considerable debate surrounds the various methods used to put the pipe through the roadbed. The crux of the controversy is whether or not there is any advantage to placing the carrier pipe (the pipeline itself) inside a larger pipe, known as *casing*. The casing is intended to increase the safety of the carrier pipe by shielding it from the greater load stress of vehicular traffic. Casing is mandatory when crossing under railroad right-of-ways. In recent years, casing has been used less and less because of improvements in corrosion control and cathodic protection and the development of higher-grade steels and better welding techniques. The use of casing is still mandatory, however, in certain locations, and these are usually identified in state and local regulations.

Figure 4.27. Pipelines are bored beneath the roadbed to avoid disrupting traffic.

ROAD CROSSINGS
WITH CASINGS

The machinery used to cross beneath the roadbed is designed to bore the hole for the pipeline while pushing the casing through at the same time (fig. 4.28). A side boom supports the boring machine and its operator and keeps them in correct position. An auger shaft extending from the machine to the far end of the casing has a cutting head, or auger, that rotates just inside the casing. As it moves forward, it bores a hole big enough for the casing to slip through. To prevent undermining the roadbed, the cutting head should not extend beyond the end of the pipe. Loosened dirt falling from above also tends to clog the cutting head if there is a gap between it and the casing. Cuttings from the hole are removed by the auger. Drilling mud is often pumped through a small pipe mounted on the casing in order to lubricate the cutting head. The mud aids both in removing the cuttings and in reducing friction between the casing and the bored hole.

A pulling force used to drive the casing under the roadbed is generated by a cable run from the hydraulic winch on the boring machine to a so-called deadman set at a right angle to the ditch above the hole. The deadman serves as an anchoring point against which the winch can pull. The boring machine engine turns the auger and powers the winch. As the cutting head rotates, a constant push drives it forward. Skill and experience are required on the part of the operator to balance the two forces and thus ensure smooth boring.

Calculations are made before and during boring to maintain the alignment necessary to guide the casing under the roadbed. This precision is imperative if the contractor is to comply with the strict cover requirements for pipe under roadbeds.

After the casing exits on the other side of the roadbed, the casing is then thoroughly swabbed to remove any moisture and debris. Side booms position the carrier pipe in the casing and help reduce sliding friction and prevent damage to the pipe coating.

Figure 4.28. Supported by side booms, the boring machine pushes the casing under the roadbed.

When it has been determined that the pipe is in place and secure, the casing ends are sealed and a vent is installed at one end. Should a leak develop in the pipeline, the fluid will be forced through the vent, thus aiding in locating the break. The vent also serves to mark the boundary of the highway right-of-way.

ROAD CROSSINGS WITH UNCASED PIPE

Whenever specifications allow, road crossings are bored without casing. On uncased crossings, the carrier pipe itself is pushed under the roadbed by the road-boring machine. This joint usually has a heavier wall thickness than the rest of the line pipe so that it can withstand the greater external load factor thought to exist at these crossings. A drawback to uncased crossings is the potential for damaging the pipe coating during installation.

One solution for lessening the likelihood of coating damage is slick boring. Slick boring gets its name from the large amount of liquid that is pumped into the hole outside the pipe to reduce friction. The most commonly used lubricant is a mixture of bentonite and water that is pumped into the hole under pressure. This is especially useful in sandy soils where abrasion is a problem. With certain highly viscous soils such as clay, water alone is often an adequate lubricant.

In an attempt to further safeguard the pipe coating, some contractors use a dummy pipe when slick boring. In a way, this approach is similar to using casing. The difference is that the dummy pipe is bored through to create the hole. Then, carrier pipe and the dummy pipe are welded together. As the dummy end of the pipe is pulled through the hole, the carrier pipe is pulled into position. Once the carrier pipe is in place, the weld connecting the two joints is cut.

WET BORING

Wet boring is similar to slick boring and is occasionally used for small-diameter pipe such as that found on natural gas distribution lines (fig. 4.29). A pilot string is first pushed through the hole, with large quantities of water functioning as lubricant. After the pilot string breaks through on the opposite side, a cutting head and the carrier pipe are attached to it. As the string is drawn back through toward its original starting point, the cutting head is rotated to clear a hole large enough to fit the carrier pipe pulled behind it. A swivel attachment behind the cutting head allows it to rotate while the carrier pipe is kept stationary.

Figure 4.29. Drilling a pilot hole for a wet-boring operation

OTHER METHODS USED FOR ROAD CROSSINGS

Although road boring is by far the most popular method for road crossings, there are several situations for which it is unsuited. One example is a roadbed with a solid rock base. For obvious reasons, explosives cannot be used. The contractor's alternative is to tunnel through the rock. It should be emphasized that tunneling is extremely expensive and thus is rarely done. Just as with the ditching machine in solid rock, newly developed boring equipment such as conical rock heads and boring machines with variable speed transmissions have enabled contractors to bore through materials that previously would have required tunneling.

The difference between tunneling and boring is significant. Boring is an extensively mechanized procedure. Most or all of the work is done by machine, and only rarely is manual excavation involved. Tunneling, on the other hand is predominantly a manual operation. It is comparable in many ways to work in the early coal mines. The tunnel itself is relatively small—just large enough for one person to work inside. Not surprisingly, the work is tedious, and the progress is extremely slow.

In sparsely populated areas, the right-of-way often crosses little-used dirt or gravel roads. In these instances, the stipulation that crossings must be bored under all roads carrying vehicular traffic is sometimes waived, and the normal pipe-laying sequence is followed with the ditch cut through the road itself. Unlike tunnels, which are necessary to overcome a physical obstacle, open-cut crossings are made for convenience and to hold down costs.

Welding

The importance of welding and its attendant functions cannot be overemphasized. Welding is the pivotal operation in pipeline construction, and the proficiency with which it is accomplished largely determines the quality, safety, and operational life of the finished pipeline. In addition, the completion of the pipeline directly depends on the speed at which the welding crews work.

To ensure optimal welding conditions and productivity, the pipe must be set up properly. Proper pipe setup involves both correct positioning along the right-of-way and, most important, careful preparation of the pipe ends. The pipe gang is responsible for positioning the pipe, aligning it, and making the initial welds. The pipe gang could rightly be called the heartbeat of the spread. Working quickly but with precision, the pipe gang sets the pace that will determine the progress of the rest of the spread. The fluid movements of an experienced pipe gang make the job appear deceptively simple. In fact, highly trained, skilled hands and eyes are needed to achieve the precise tolerances required for effective alignment and welding (fig. 4.30).

Before the pipe can be welded, however, the pipe ends must be thoroughly cleaned of any dirt, rust, mill scale, or solvent. Power hand tools, such as wire brushes and buffers, are still the choice for completing these grinding and buffing tasks. The

Figure 4.31. Grinding pipe ends prior to welding

time and care taken during this initial cleaning is cheap insurance against flawed welds that may later require a costly and time-consuming repair job (fig. 4.31).

Before the pipe is delivered to the job site, the ends are cut, or beveled, to an angle that will best accommodate the welding technique being used on the job. A commonly used bevel angle is 30 degrees. Variations are possible depending on the welding process used, the grade of steel, and the operator's specifications.

While the majority of beveling work is done at the mill before the pipe is delivered to the job site, there are circumstances that require rebeveling in the field. This most often occurs when a defective weld must be cut out of the line or when the pipe end has been damaged during transport and handling. Field beveling is done in one of several ways. One method involves using an end prep, or pipe-facing, machine to produce a precise angle, or shoulder, in a very short time. A second technique achieves the same results but uses a small torch to cut the bevel in the pipe end.

PREHEATING

Preheating the pipe ends before welding is not always required but has become standard practice in certain situations. Ambient temperatures below 40°F, for example, or overnight condensation of moisture on the pipe usually make preheating necessary. In addition, high alloy or carbon content found in certain grades of steel and especially on heavy-wall, large-diameter pipe, also call for preheating.

Standard procedure calls for heating 3 to 4 inches of pipe behind the bevel to 200°F. To maintain this temperature or to raise the temperature to the required level, wagon-wheel heaters are often used. These multiheaded circular propane torches are manually rotated inside the end of the pipe to facilitate even temperature distribution (fig. 4.32).

ALIGNMENT

Perhaps the true test of a good pipe gang is its ability to align pipe joints quickly and accurately. Speed is certainly essential to the operation, but a certain deftness of touch born of long experience is also indispensable. To begin the procedure, the side-boom operator raises and supports the joint of pipe and then maneuvers it into a position level with the pipe string already in place. The side-boom operator must also have exceptional skill and dexterity in order to gently position a joint of pipe that may weigh as much as 10 tons. As the new joint is maneuvered into place, an internal line-up clamp operated by means of a long steel rod is run through the last joint to the point where the pipe ends meet. An air hose that supplies pneumatic power to the clamp is also run alongside the steel rod inside the pipe joint. The line-up clamp uses a number of small expandable blocks, or shoes, to grip the inside surfaces of both pipe joints and hold them in place (fig. 4.33). Minor out-of-round deformities can also be corrected during alignment, particularly with large-diameter, thin-walled pipe that tends to flex more and is thus more susceptible to minor disruptions. The internal clamp can also act as a swab to clean the inside of the pipe, thus removing any debris that may have accumulated. Certain special situations may

Figure 4.32. Preheating the pipe ends

Figure 4.33. To prepare for welding, internal line-up clamps are inserted and the joints are then positioned and aligned.

call for an external clamp instead of the internal clamp. External clamps are usually used on pipe with a diameter of 8 inches or less (fig. 4.34).

An important consideration during alignment is the spacing of the narrow width between the bevels of the adjoining pipe ends. The welding process to be used determines the width of this gap, which is crucial to adequate penetration of the weld material into the space between the bevels. Highly skilled members of the pipe gang, known as *spacers,* are charged with this responsibility. The spacers strike a wedge into the interface between the pipe bevels and then maneuver them to an exact, uniform distance around the entire circumference. This spacing must be maintained on all joints throughout the line.

After these preliminary tasks are finished, the pipe is ready for the initial welds. Without question, the most

important part of pipeline construction has always been the ability to make field girth welds at high productivity rates while maintaining high quality to avoid costly weld repairs and cutouts.

The initial weld is referred to as the *root bead.* The root bead is made while the internal line-up clamp is still in place. Two welders work on opposite sides of the pipe, and each is responsible for welding his portion of the pipe circumference (fig. 4.35). Pressure on the inside clamp is maintained until the root bead is finished. The clamp is then carefully released and withdrawn to start the next joint. Any movement or undue stress on the fresh weld must be avoided. After the correct alignment has been achieved, a partial joint is welded, and wooden skids are placed under the pipe end to support it. This preliminary welding allows the pipe gang to move ahead and align the next joint.

Figure 4.34. While more commonly used on small-diameter pipe, special situations may require the use of different types of external clamps.

The pipe gang is actually responsible for two welding operations. One crew makes the root bead, also called the *stringer bead* (fig. 4.36). After release of the line-up clamp, a second crew follows immediately behind the stringer bead crew and makes the hot-pass weld (fig. 4.37). The wall thickness of the pipe dictates the number of hot-pass beads that are necessary.

Figure 4.35. Welders work on opposite sides of the pipe.

Figure 4.36. Stringer bead

Figure 4.37. Hot pass

54

Following the completion of the root bead and hot-pass welds, another welding crew—the firing line—takes over. The firing line builds on the weld material already in place. These additional welds, known as *filler beads*, are usually made continuously over the previous weld (fig. 4.38). When welding is interrupted for any length of time, preheating may be necessary before welding can resume. Care must be taken to grind and clean the surface of each weld before the next bead is made. The final welding pass, called the *cap bead,* completes the joint.

Welding crews can use any one of several different welding techniques to join pipe ends. Among those possible are manual welding, wire welding (usually semiautomatic), a combination of these two, and automatic welding. Variables such as the grade of steel used in the manufacture of the pipe, weather conditions, and terrain all have a bearing on the final choice. It is quite common to combine the manual and wire processes, especially for large-diameter pipe. Wire welding is used for making the root bead, and the filler beads and the cap bead are welded manually.

Figure 4.38. Firing line

Figure 4.39. Submerged-arc welding process

MANUAL WELDING

Manual, or stick, welding produces an electric arc that melts and fuses the pipe ends with the metal of the stick electrode held by the welder. An electric current source is supplied by portable welding generators mounted either on a welding tractor or on the individual welder's truck or rig. The stick electrodes are metal rods generally $\frac{5}{32}$ or $\frac{3}{16}$ inch in diameter and from 9 to 18 inches in length. Until the early 1930s, welding rods consisted of nothing more than pieces of bare metal. As welding technology progressed, coatings were applied to the rods. At high temperatures, these coatings either create an inert gas shield or react with free oxygen. In either case, the weld is protected from rapid oxidation caused by contact with the oxygen in the atmosphere. This shielded-arc process results in welds that are extremely fine grained, are free of oxides and nitrides, and have great ductility and toughness. The rod coating also forms a slag deposit on the pool of molten metal. This deposit, which hardens into a protective crust after the weld cools, must be ground away before the next welding pass.

SEMIAUTOMATIC WELDING

With semiautomatic welding, the arc is maintained in a continuous stream of gas between an electrode and the pipe. The arc heats the pipe and melts the electrode, thus supplying filler metal for the joint. The welding apparatus consists of a spool of coiled wire, a pair of driving rolls, a welding gun incorporating a control switch, and a gas supply. The particular type most frequently used on pipeline work is called CO_2 wire welding because carbon dioxide is the gas used to shield the welding process. The wire used is usually 0.035 or 0.045 inch in diameter. Since the rate of wire feed and the flow of gas are controlled by the welder, the process is semiautomatic. Semiautomatic welding is frequently used on the stringer-bead pass because the weld produced by this method has a high tensile strength. Also, the rate of metal deposition is higher with this method than it is with stick welding, thus making it popular for heavy-wall pipe.

AUTOMATIC WELDING

Two general types of automatic welding are used on pipeline construction work: submerged-arc welding and automatic wire welding. The submerged-arc welding process utilizes a continuous wire feed and a shielding medium of fusible granular flux (fig. 4.39). This process is known for its high deposition rates and weld passes of substantial thickness. It is mainly used in double-joint racks at pipe storage yards or coating plants, where 40-foot joints are welded together into 80-foot joints. The high welding speed and good weld quality make the submerged-arc welding process very economical.

The extensive use of automatic welding in mill and fabrication work has found ready adaptation in the field for making welds on the large-diameter, heavy-wall pipe now commonly used. By definition, automatic welding implies a technique that minimizes the necessity of constant manual adjustment and control of the welding parameters.

Under certain conditions, this automation can be extremely advantageous. High production rates and overall high quality are possible with this type of welding because an automatic welding head operates in a narrow welding groove. The narrow groove allows less weld material to be used, and the amount that is needed is laid down at a fast rate and at a high voltage and amperage setting.

Several different types of automatic welding setups are available to the contractor (fig. 4.40). All of them operate on essentially the same basic principles. The differences among the techniques usually involve the type of wire and shielding gas used and the mounting frame for the welder. The shielding medium is an inert gas, and its primary purpose is to prevent oxidation of the weld at the point of contact with the pipe metal. This is accomplished by excluding oxygen in the air from the area around the molten metal. In addition, protection from crosswinds that might disperse the gas shield is provided by a canvas canopy over the bead heads and pipe joint.

Shielding gases can be combined and adjusted to varying percentages to suit the wall thickness and the grade of steel of the pipe. Manipulation of the gas shield affords control of the depth and penetration of the weld arc,

Figure 4.40. Automatic welding setup

thereby substantially reducing the chance of a burnthrough. Argon and helium are good for this purpose, but in the field, carbon dioxide is the most widely used shielding gas. In all cases where full penetration is desired, protection from molten metal on the inside surface of the pipe is necessary.

WELDING QUALITY CONTROL

Safety and quality control are the paramount concerns in every aspect of pipeline construction. Constant inspection, testing, and monitoring ensure the safety of those working on the pipeline as well as the operating integrity of the pipeline itself. Consistent with the priority placed on welding, quality control measures are especially stringent in this area.

Much of the credit for the consistently high safety record of pipelines in general can be attributed to the progress made in welding inspection and testing. In the early days the industry relied on simple visual examinations to detect flaws in the weld. The introduction of radiographic inspection techniques in 1948 proved to be a milestone. Since that time equally important improvements have been made.

The acceptability of welds can be determined by either of two different testing methods, namely nondestructive or destructive testing. As the term implies, nondestructive tests are those designed to evaluate the quality of both production and field welds without altering their basic properties or affecting their future usefulness. Currently the most widely used nondestructive technique is radiographic, or X-ray, testing.

X rays provide an accurate and relatively quick means of examining a weld for possible defects. X rays are directed through the weld and onto radiographic film. The developed film produces a shadowgraph of the weld and clearly highlights any flaws. Both X-ray machines and radioisotopes are utilized as sources for X rays. Regardless of the source used, strict safety precautions are observed to guard personnel against accidental overexposure to harmful radiation.

Self-propelled X-ray machines, or crawlers, are now available. These crawlers ride inside the pipe and can be programmed to automatically stop at specific joints. In this way X rays may be taken of every joint, every other joint, or any such sequence desired (fig. 4.41).

X-ray film is developed on the job site in portable darkrooms so that any needed corrective measures can be taken immediately. Standards vary according to the contract specifications and government regulations, but most jobs now require X rays to be taken of 100 percent of the welds made. This is particularly true for pipelines passing through or close to population centers or environmentally sensitive areas. X rays are normally kept on file for at least three years after the pipeline is completed. While X-ray testing is still the preferred method used on most projects, techniques such as magnetic particle tracing, ultrasonics, and fracture mechanics are also gaining acceptance.

In contrast to nondestructive testing, destructive testing literally tears the weld apart to examine its structure. It is primarily used during qualification procedures required of

Figure 4.41. An X-ray crawler is used to determine weld integrity.

all welders who work on a pipeline. Pulling machines are used to stress the weld to the breaking point while measuring the amount of force required to do it. Destructive testing is occasionally used in the field to test the quality of welds.

Unlike most other welders, those working on a pipeline are required to pass a series of proficiency tests. Pipeline work involves making welds in positions other than the simple horizontal ones common to many kinds of welding work. In pipeline welding the pool of molten metal must be manipulated from flat to vertical to overhead, and skills significantly more advanced than those needed on other welding jobs are demanded. Acceptable performance on qualifying procedures is judged in accordance with guidelines set out in API Standard 1104. Under this standard, welds may be subjected to tests including a tensile test, a bend test to measure ductility, a Charpy test to determine the energy required to break the weld, a nick-break test to measure hardness, and X-ray examination. Testing can be done in the field with a portable pulling machine.

Cathodic Protection

Corrosion is the archenemy of any buried metallic structure. In the early 1980s in the United States alone, the annual loss due to corrision was estimated to be some $70 billion when both the energy cost and the replacement price of the damaged structures were added together. In the case of pipelines, this cost also takes into account the extra energy required to push the pipeline fluids past corrosion obstructions.

Corrosion is essentially electrochemical in nature. Most susceptible to destruction is a metal or alloy that is immersed in an electrically ionized, conductive environment. In pipeline work, soil moisture and chemicals leaked from the soil and held in solution in groundwater provide an ideal medium for this electrochemical reaction to take place. The problem is intensified in pipelines submerged offshore, in swamps, or in river beds. A number of design factors also contribute to the formation of corrosion. For example, thermal stress, which is generated both by high operating temperatures and by cyclical high and low temperatures, is found most often in pipelines subject to frequent start-ups and shutdowns. Regardless of the source or cause of corrosion, certain standard precautions are now employed to prevent a corrosion cell from forming.

During the corrosion process, an electrochemical potential that allows current to flow from an area of greater potential (the anode) to one of lesser potential (the cathode) is generated. The anode—in this case the buried pipe—is eventually corroded and destroyed as current flows away from it and the metal oxidizes. If, however, a situation is created where a current equal to that being lost through corrosion is directed toward the pipe, the process can be reversed, and corrosion can be effectively inhibited. Today's pipeline engineers place carbon anode current sources in the soil in order to maintain a current balance and enhance corrosion protection. This process, called cathodic protection, does not necessarily eliminate corrosion. Rather, it acts to deflect potentially corrosive influences away from the structure being protected and concentrates them at another known location.

Pipe Coatings

Cathodic protection plays just a small part in protecting the line from corrosion. Coating and wrapping pipe with special sealing materials is the main line of defense against corrosion. The primary function of pipe coatings is to prevent water from coming into contact with the steel of the pipe. Thus, the most important thing to consider when a pipe coating is being selected is its ability to resist water penetration. This is important, of course, because absorbed moisture can carry electrical current through the coating material itself and onto the pipe, thereby initiating the corrosion process.

The moisture content of the soil as well as a number of other external factors must be considered when selecting a pipe coating. The location of the pipeline in relation to buildings, power lines, population centers, and other pipelines may mean that a tougher, more durable coating is required in order to comply with safety regulations governing such sites. The qualities of the soil in which the line is buried may also dictate a specialized coating. For example, hydrocarbon contamination in the soil hastens the degradation of some pipe coatings, and the abrasive effects of soils that are alternately wet and dry can also stress some coatings to the point of failure. Temperature also can affect pipe coatings. Just as a pipeline's operating temperature may lead to corrosion problems, the prevailing temperature during construction and installation of the line can also present difficulties. Extremely hot or cold conditions during construction can ultimately affect the operational life of the coating by causing thermal expansion and contraction of the pipe, which stress the pipe coating and interfere with bonding.

Other things that affect the longevity and integrity of pipe coatings are the conditions under which the coated pipe is stored and the handling the pipe receives during lowering-in. Pipe with mill- or factory-applied coatings must be carefully stored to minimize coating distortion and damage. While damage most often results from careless handling, allowing pipe to stand unprotected in a storage yard may promote undue thermal stress on the coating. For this reason, whitewash, kraft paper, or a sacrificial layer of tape is used to protect pipe that is kept outdoors for any length of time. These coverings reflect ultraviolet light and dissipate any heat buildup so that pipe coatings do not soften at high temperatures. Applying whitewash can effectively lower the temperature of pipe by as much as 60°F.

Whether a coating is applied at the mill or on the right-of-way, it is most vulnerable to damage when the pipe is being lowered in. No matter how carefully this procedure is carried out, some damage is inevitable, and a wrapping is often placed over the pipe coating to provide additional mechanical support. Some of the materials used for wrapping are felt, fiberglass, fiberglass-reinforced felt, and kraft paper. Kraft paper can serve a dual purpose by both deflecting ultraviolet rays and providing a ready indicator of any abrasion or puncture of the coating.

A common type of coating failure is disbonding, or separation, of the coating from the pipe. The problem can usually be traced to inadequate preparation of the pipe surface before the coating was applied. In some cases, however, the pressure differential created by a voltage imbalance between the pipe surface and the soil electrolyte can build to the point where water is actually forced to migrate through the coating and onto the surface of the pipe. Disbonding of the coating and the beginning of a corrosion cell is the usual result of this process.

TYPES OF COATINGS

Different types of coatings are available to meet the contractual demands set out by the pipeline owners as well as to accommodate the preferences of individual contractors. The most widely used types are bituminous enamels, epoxy resins, and tapes.

Enamel Coatings. Bituminous enamels are derived from coal-tar pitches or petroleum asphalts.

Because of their wide availability, low cost, and ease of application, they have been widely used throughout the industry almost since its inception. Enamel coatings include a wide variety of petroleum-based derivatives such as asphalts, coal tars, greases and waxes, mastics, and asphalt mastics. Asphalts are substances that occur both naturally and as a byproduct of petroleum distillation, while coal tar is a mixture of tars formulated from the byproducts of coal distillation to which various filler substances have been added. Grease and wax are petroleum-based coatings containing corrosion inhibitors and fillers. The use of grease and wax has declined markedly, and they are only rarely used in modern pipeline construction. Mastics are a mixture of asphalt, sand, and miscellaneous fillers. They are used when a thicker, more durable consistency and greater adhesion are needed. Asphalt mastic, a form of coating that has been in use in the industry for over 50 years, is a dense mixture of sand, crushed limestone, and fiber bound together with asphalt. Because it is the thickest of all the corrosion coatings, it is generally used to greatest advantage in offshore pipelines. Both asphalt mastics and the other enamel coatings are not suitable for use in areas where the soil has been contaminated with hydrocarbons.

With enamel coatings, the material in intimate contact with the pipe surface is the enamel. The sand, felt, fiberglass, and other materials used as fillers function as mechanical reinforcement to protect the pipe during construction. Both hot- and cold-applied enamels are available, and which is used depends on the requirements of the job and contract specifications. Enamel coatings are most effective when the pipeline's operating temperature range is between 30°F and 180°F. Below 40°F, however, precautions must be taken to guard against cracking and subsequent bonding problems.

Fusion-bonded Epoxy Coatings. Fusion-bonded epoxies are powdered resins that form a virtual skin over the steel surface when applied to heated pipe. They are usually applied at the mill because of the specialized equipment needed and the exacting preapplication requirements (fig. 4.42).

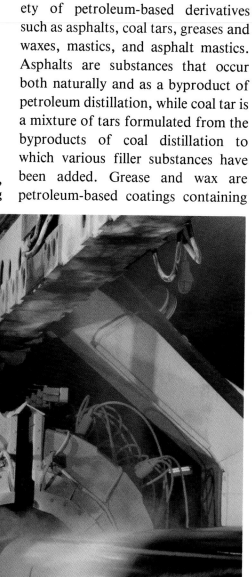

Figure 4.42. Mill-applied, fusion-bonded epoxy coating

For maximum effectiveness and bonding strength, pipe surfaces must be blast-cleaned to a near-white finish. The entire pipe is then heated to a specified temperature, usually between 400°F and 500°F, and a fine spray of powdered epoxy resin is directed onto the pipe surface. The resin melts on contact, and the residual heat of the pipe cures the coating and bonds it to the pipe.

One of the principal advantages of the fusion-bonded epoxies is the relatively thin coating they leave. Because defects are not hidden under a thick layer of coating or outer wrap, as is often the case with other methods, final inspection of the pipe surface is considerably easier. Other advantages include the coating's strong resistance to handling damage and its overall durability and flexibility. This latter characteristic also allows epoxy-coated pipe to be bent in the field with little danger of damaging the coating. Another advantage is that fusion-bonded epoxies can also be applied to the internal surfaces of pipelines. Gas transmission lines in particular are more susceptible to internal corrosion because of the hydrogen sulfide, carbon dioxide, and water found in natural gas. Fusion-bonded epoxies can also withstand the high operating temperatures (200°F or higher) commonly found in gas transmission lines.

Tape Coatings. Polyethylene tape is now being used more and more for line travel applied external pipe coatings. Although polyethylene is the most commonly used tape coating, several other types are also available. These include polyvinyl, coal tar base, and butyl-mastic adhesive coatings. Thicker tapes and stronger resins are

becoming more popular. New adhesives and primers that offer greater bonding strength are also being developed. Since the coating is applied on the line, its integrity can be immediately verified, and any defects can be repaired relatively quickly and simply. Another advantage is that the application procedure does not raise any serious environmental or health objections.

LINE TRAVEL APPLIED COATINGS

While the type of coating and the method of application selected are instrumental in combatting pipe corrosion, their ultimate success depends on the preparation of the pipe itself. For line travel applied coatings the pipe surface must be free of all dirt, mill scale, rust, and debris. A self-propelled cleaning and priming machine with a rotating set of brushes and buffers rides the top of the pipe, removing any loose material (fig. 4.43). At the same time, a thin coat of quick-drying primer is applied to prepare the pipe for coating.

Figure 4.43. Self-propelled cleaning and priming machine

After the cleaning and priming process has been completed, a coating and wrapping machine begins to move immediately down the prepared pipe (fig. 4.44), minimizing contamination of the newly cleaned pipe. When line travel tape is used, the cleaning and priming and the coating and wrapping processes are performed with one machine (fig. 4.45). As many as five side booms may be necessary for line travel applied coatings. The first tractors lift the pipe to provide clearance for the tape or coating machine. Then, another tractor supports the coating machine, while the last tractor lifts the back end of the pipe section and tows a sled containing coating materials and supplies.

Coal tar enamels are a particularly effective coating that can be applied over the ditch. The enamel is brought to the right-of-way in large blocks that are melted in dope pots (fig. 4.46). These portable containers heat the enamel to the temperature required for application and then maintain that temperature throughout the coating operation. Agitator paddles constantly stir the mixture to ensure even distribution of heat and to prevent the breakdown of coating elements that can occur if any part of the enamel remains in contact too long with the heat source at the bottom of the kettle. The heated liquid enamel is fed to the coating machine through a flexible hose connection, and an even layer is applied to the pipe surface. The coating is then wrapped with fiberglass, felt, and kraft paper to protect it

Figure 4.44. Hot-enamel coating machine

Figure 4.46. Firing the dope pots

Figure 4.45. Clean, prime, and tape machine

both while it hardens and while the pipe is being lowered in (fig. 4.47). The taping operation is essentially similar, with the obvious absence of the dope kettle. As the taping machine moves along the pipe, tape is wrapped around the pipe in overlapping segments (fig. 4.48). Rolls of tape are mounted on spindles attached to a rotating wheel on the rear of the tape machine. Adjustable arms maintain the necessary tension to prevent slack, wrinkling, or missed spaces, particularly when taping in rough or uneven terrain, where maintaining tension may be more difficult.

Figure 4.47. Hot dope gang

Figure 4.48. Tape crew moving down the right-of-way

On pipe that has been mill coated, a cutback of 6 to 9 inches is usually left at the end of the joint so that the coating does not interfere with the welding process. After the weld is completed, these field joints must be coated. A variety of processes or materials may be used, depending on the type of coating applied at the mill. For example, coal tar, glass, and felt are usually used on mill-coated coal tar pipe. Heat-applied epoxy powder can be used with thin-film coatings. If the coatings are compatible, however, almost any type of coating can be used on a field joint. Before the coating process is begun, the pipe surface must be cleaned by power brushing or sand blasting. A primer is then applied, followed by the selected coating (fig. 4.49). The most important consideration is a proper bond with the adjacent coating to ensure a waterproof joint.

Figure 4.49. Three types of coating for field joints: *A,* **dope;** *B,* **tape;** *C,* **powdered epoxy**

A

B

C

DETECTING COATING FLAWS

Machine Detection. Detection of coating flaws has evolved into a science in its own right and is an important quality check. The basic principle behind this technique involves establishing an electric potential between the pipe and an external electrode with the help of a machine called a *holiday detector,* or jeep. It consists of a metallic ring that encircles the pipe, a hand-held voltage source, and an alarm unit. The amount of voltage applied to the ring can be adjusted according to the type of coating. Thicker coatings, such as coal tars, generally require a higher voltage setting. Too much voltage, however, can damage the thinner coatings, such as tapes. The operator pushes or rolls the holiday detector along the pipe (fig. 4.50). When the detector passes over a gap, or holiday, in the coating, an electrical spark completes the circuit between the electrode and the pipe. An alarm or buzzer on the unit is then activated, and the exact location of the holiday is marked. Flaws discovered in this manner are repaired manually.

A coating flaw, or holiday, can be repaired in several different ways. It can be patched by removing the coating around the holiday and then recoating that small area. Another method is to recoat the entire circumference of the pipe, thus eliminating the defect. Regardless of the method chosen, great care must be taken to ensure a proper bond with the surrounding coating so as to eliminate any potential water entry.

A different type of holiday detector checks for coating defects as well as any metal debris near the line that may cause future corrosion problems. Named for its inventor, the Pearson holiday detector is passed over the line after the pipe has been lowered into the ditch and buried. An excellent additional safety check, the Pearson could be regarded as a sort of glorified mine detector, since the underlying operating principles of the two devices are essentially the same. Although digging the pipe back up to repair a coating flaw or remove some debris is expensive, it is necessary if greater problems and expense in the form of corrosion and pipe failure are to be avoided later.

Visual Inspection. While sophisticated techniques such as those previously described have taken much of the guesswork out of locating coating flaws, there is still a role for on-site, visual inspection. An experienced eye, which can often detect a problem or prevent one from occurring, is the ultimate quality control measure.

Figure 4.50. Jeeping the pipe coating with a holiday detector

Lowering-In

Pipe can be lowered directly into the ditch as part of the coating operation or by a separate lowering-in crew if the coating has to set up before lowering or if the pipe has been mill coated. It must be remembered, however, that any time the pipe is moved after it has been coated, there is a possibility of damaging the coating or the pipe itself. This risk is compounded when the pipe is being lowered in, and successful lowering-in relies heavily on the deftness of the side-boom operators who handle the pipe. Considerable coordination and timing are needed when two or more machines are being used to move heavy sections of pipe (fig. 4.51). Finesse rather than sheer power is the decisive element in placing the pipe safely into the ditch. Using nonmetallic slings or belts suspended from the side booms, the pipe is lifted evenly.

Correct spacing of these slings lessens the chance of pressure damage to the coating and helps maintain proper weight distribution. Too great a section of unsupported pipe leads to buckling and other damage.

To help counter the effects of pipe contraction and expansion due to extreme daily temperature differences, pipe can be roped into the ditch to create slack loops. They are made by alternately laying pipe on opposite sides of the ditch. When used at all, slack loops are more common for smaller-diameter pipe. This technique has drawbacks when used with big-inch pipe because of the weight and size of the pipe joints. In modern pipeline construction, concern over the possibility of temperature changes producing tension and compression of the pipe has been considerably lessened largely because of advances in metallurgy and pipeline construction techniques.

Figure 4.51. Two views of lowering-in

Once in the ditch, movement of pipe joints should be kept to an absolute minimum. A small adjustment may occasionally be necessary, but anything beyond that invites trouble. Damage or defects in workmanship usually mean a long delay and considerable expense to correct. These consequences should be kept in mind when handling the pipe this final time.

Backfilling

After pipe has been lowered into the ditch, it requires more than just a random burial. Instead, adequate fill material must be provided underneath the pipe as well as above it. In the long run, systematic backfilling combined with other auxiliary measures can help prevent pipe damage due to loose rock, abrasion, shifting, and washouts.

A wide range of equipment is used for backfilling, and selection of the most suitable depends on the type of fill material and the condition of the ditch. If a ditch bottom is hard and flat, for example, it is more difficult to place dirt so that it reaches the lower quadrants of the pipe. As settling occurs, the weight of both the pipe and the backfill is supported by only a narrow strip along the ditch bottom. Potentially damaging shear stresses may then develop, increasing the likelihood of coating damage and pipe failure. An auger fitted to the front of a bulldozer can often alleviate this problem (fig. 4.52). The corkscrew action of the auger blade churns the fill evenly into the ditch.

The U.S. Department of Transportation has established a regulatory code that specifies minimum ditch depth and cover requirements. Pipelines are classified according to the diameter of the pipe, location of the line, and the type of material to be used as backfill. For example, a line may need only 30 inches of backfill when the right-of-way passes through an isolated rural area. Upon entering a town, however, cover requirements may be increased by an additional 6 to 18 inches. In all cases, cover depth is measured from the top of the pipe to ground level along the right-of-way.

Figure 4.52. An auger distributes the backfill evenly in the ditch.

Pipe laid in rocky or rough soils usually requires padding on the ditch bottom. Dirt used for this purpose should be screened or sifted to eliminate rocks and large chunks. Pipe may also be laid on earth-filled sacks (fig. 4.53). Synthetic materials such as foam can also provide a uniform, padded surface on the ditch bottom. When placed along the upper surface of the pipe, foam is also an effective rock shield (fig. 4.54). Pipelines carrying products that must remain within a specific temperature range have also been successfully insulated with foam.

Where ditching has been accomplished by blasting through solid rock, the spoil that remains is not acceptable as backfill. Dirt or other material must be trucked in to pad the ditch bottom and to fill most of the ditch. Some contracts specify that only the first several feet be clean fill. The remainder of the ditch can be filled with small rock to within 1 foot of the top. A mound or crown of clean fill is then added as the final layer.

Backfilling frozen soil presents problems similar to those found with rock. If it is not broken up before being returned to the ditch, frozen soil can damage pipe coatings because of the irregular shapes of the frozen pieces. Also, frozen soil used as fill does not distribute itself evenly. Uneven distribution may leave sections of the pipe unsupported after the fill has thawed. Such free spans are prone to abnormal stress and are more likely to develop cracks and leaks.

Earth that is frozen solid on the surface may be loose when dug below the frost line. If the spoil bank is left to stand, however, the moisture in the soil will freeze and solidify the dirt. In some areas, mild daytime temperatures can drop drastically at night, converting a loose spoil bank into rock

Figure 4.53. Padding the ditch is sometimes necessary to prevent coating damage.

Figure 4.54. Foam can serve many purposes in protecting the pipe.

overnight. Where such temperature fluctuations are common, it is often desirable to open and close the ditch on the same day. Heavy equipment can also be used to try to break up frozen backfill, but by the time such measures are necessary it is probably already too late.

While pipe shifting is usually attributable to the steel's reaction to temperature changes, a hard rain for several days or a sudden heavy downpour can literally float the pipe in the ditch. If the rate of water flow is rapid enough, the stress on the pipe can be severe, and special anchoring devices or ditch breakers may have to be installed on the pipe or in the ditch to prevent problems.

Anchoring devices secure the pipe in the ditch at stream crossings, in swamps, and in other areas likely to flood. Because of the force that moving water can generate, concrete anchors weighing 5 tons or more are routinely used and these may be spaced as close as every 10 feet (fig. 4.55). Two types of anchors are commonly used. Saddle-type concrete anchors literally straddle the pipe in order to hold it down, while bolt-on anchors cast from concrete encircle the pipe at carefully calculated intervals to ensure stability. Other devices, such as screw anchors, are also available, and they have holding capacities rated up to 18 tons when correctly applied. Where there is the threat of washouts, ditch breakers are used to form internal barriers to water movement. Sandbags and, more recently, foam partitions are employed to divide the ditch into sections to maintain the structure of the ditch and to secure the pipe.

After the ditch has been filled, any topsoil that initially was stripped off is replaced. Where erosion prevention is a consideration, the backfill can be compacted by running a bulldozer or other heavy, tracked machinery along the ditch. In most cases compacting is unnecessary, since settling and compaction occur normally.

Figure 4.55. Concrete weights are used to anchor pipe.

Tie-In

A pipeline spread has been compared to an assembly line where the workers move instead of the components. Although a considerable distance often separates each crew, an overall rhythm is apparent. Ideally, the spread maintains this rhythm, and progress is steady from beginning to end. To avoid a loss of rhythm, situations that might cause a break in the continuity of pipe-laying operations are routinely bypassed by the regular crews. This work is later completed by the tie-in gang (fig. 4.56).

Tie-in mainly involves welding road and river crossings, valves, portions of the pipeline left disconnected for hydrostatic testing, and other special fabrications the pipeline may require. Any additional work, such as taping or coating the welds, is also done by this crew. The tie-in gang essentially functions as a self-contained spread, and their sole purpose is to attend to work left behind.

When it is known that a tie-in will be required at a particular point, a carefully measured extra length is left on the adjoining pipe ends. Because of the exact tolerances involved, pipe ends are cut with a beveling machine. The cutting torch head is mounted on a ring attached to the outside of the pipe. As the head rotates on the ring around the circumference of the pipe, a perfect cut results. The contract often specifies that the remaining piece of pipe be welded into the line ahead. These pieces, called pup joints, must usually be a minimum of 10 feet in length.

Since backfilling has left just enough space to maneuver the pipe ends into position, alignment of the pipe ends is slightly more complicated than usual because of the reduced working area. For this reason, it is critical that the pipe be cut exactly to size in order to minimize the movement of pipe already buried. Excessive

Figure 4.56. Two views of the tie-in gang at work

Figure 4.57. Hot tie-in of a gas line

movement could easily scrape or gouge the coating and cause a costly holiday. Tie-in welds generally use external line-up clamps held in place until 100 percent of the stringer bead is completed. As with all other welds, those made during tie-in must be thoroughly inspected for defects.

Occasionally, tie-in welds must be made to connect pipe with lines that are already in service. When natural gas lines are connected, hot tie-ins are made. In the hot tie-in procedure, the gas in the line is purposely ignited at the point where the welding is to be done (fig. 4.57). Igniting the gas eliminates the chance of a spark that could cause an explosion of the gas and air in the line. Obviously, special precautions are necessary while working around the burning gas. These are minor inconveniences, however, compared to the potential destruction that might occur were they not taken.

Fabrication

An operating pipeline consists of more than miles and miles of relatively straight pipe. If it is to deliver products to various points along its route — while maintaining safe and efficient levels of operation — a pipeline must have a number of specialized connections and fittings, which are collectively referred to as fabrication. Fabrication assemblies are made up ahead of time and then lowered as single units directly into the ditch where they are joined to the carrier pipe.

In general, most connections of this sort serve one of three functions. The most familiar is to control product flow and direct it to the proper location. Examples of flow-control connections include mainline valves (ball, gate, butterfly, and bypass) and side taps (fig. 4.58). Other fabrications, while not as numerous or readily recognizable as valves, serve the equally important functions of maintenance and product separation. Scraper traps and launchers are used to send out and retrieve pigs and spheres. In addition to their cleaning function, pigs relay important information regarding the internal condition of the pipeline. Spheres can also be used for pipeline maintenance, or they can serve to separate different products or product grades, for example, kerosine from jet fuel. Drip settings are welded into low points in natural gas lines to trap potentially corrosive water as it condenses out of the gas stream.

A fabrication crew that works independently from the rest of the spread is responsible for welding fabrication assemblies into the line. Fabrication units are usually pretested in the fabrication yard before they are installed in the line.

Figure 4.58. Two types of valves used for flow-control connections

Cleanup and Right-of-Way Restoration

The best expression of the contractor's concern for the environment is the meticulous care he takes to clean up and restore the right-of-way. While the finished pipeline is a formidable tribute to the engineering ability of pipeliners, it will always remain hidden. With the intense interest in environmental issues, however, right-of-way restoration is an important and visible advertisement of the pipeliner's skills (fig. 4.59). It could therefore be easily argued that restoration is the most crucial step in pipelining.

The degree to which the contractor is responsible for restoring the right-of-way depends on the legal stipulation in the contract with the pipeline owner. Additional commitments may also result from agreements signed with individual landowners. Whatever the source, the principle underlying all of these obligations is the assumption that the right-of-way will be restored to its original condition or better.

Figure 4.59. Special care is taken to restore the right-of-way to its previous condition.

Figure 4.62. The fill pump is located at the water source.

Figure 4.63. The deadweight tester is used for hydrostatic testing pressure reading.

air in the line out ahead of the test water and also restricts the water to the test section. Pigs of different designs can perform a variety of functions. They can, for example, collect information on the internal condition of the pipe, clean pipe walls to promote more efficient flow, and remove any debris that may have been left in the line. Selection of the best pig for the job and following the proper procedures are critical to the test. If the pig is not sized properly or not run properly, air can become entrapped in the test section, causing problems in testing and in dewatering.

A low-head water pump set near the water source is connected to the test manifold by fill pipe (fig. 4.62). Several additional fill pumps are often required to increase volume and to boost pressure, and filters are used to clean the water before it enters the line. As the line is filled with water at a rate of several thousand gallons per minute, the pig is propelled at a rate that ensures a proper fill. The fill pipe is then disconnected, and a pressure pump is hooked up to the manifold. This pump raises the pressure of the water in the line to the specified test pressure, usually from 93 to 100 percent of the specified minimum yield strength of the line pipe.

Instruments called deadweight recorders and pressure recorders are used to monitor the pressure during the test period (fig. 4.63). Since both the ambient and the external temperatures have an effect on the test pressure, temperature recorders are also used. If the required test pressure is maintained for the proper period of time, then the test is certified and accepted. If a change in test pressure cannot be justified by corresponding

Figure 4.64. Dewatering at conclusion of hydrostatic testing

data, such as temperature fluctuation then the test pressure is maintained until the results are satisfactory or until the source of the problem can be found and repaired.

Once the test is accepted, pressure is relieved by opening the valves on the test header and reducing the pressure. Then, either compressed air supplied by the contractor or gas supplied by the owner is used to propel the pig back through the pipeline, thereby removing all the water in the section (fig. 4.64). After the water is removed, several additional pig runs may be required to dry the pipeline thoroughly. After the line is cleaned and dried, the test section is tied back into the adjacent pipeline section. The line is then ready for service.

SPECIALTY CONSTRUCTION

5

River Crossings

In pipeline construction, the quickest and most economical way around an obstacle is sometimes straight through it. This is especially true when the right-of-way intersects a river or large stream. When planning a pipeline's route, every effort is made to avoid the need for water crossings because of the expense of such construction. In most instances, however, such route alterations are either technically unfeasible or would be even more expensive than the proposed crossing itself. Contractors have become quite proficient at utilizing a variety of methods to install crossings that securely link both sides of the river without undue delay and without permanently disrupting the waterway crossed.

The most efficient method of putting in a river crossing depends on the size and type of pipeline and on the peculiarities of the river itself. Fast-flowing currents and banks that shift or recede due to erosion and flooding may eventually expose pipe that was initially thought to be adequately buried. Similarly, pipe buried in the riverbed itself may be uncovered by scouring and by the shifting of sand and soil along the bottom.

Exposed pipe is more vulnerable to failure because of accidental damage and increased stress. If the soil beneath the pipe is undermined, the unsupported span must then withstand both the direct pressure of the current and the weight of the debris that inevitably builds up around it. Marine organisms attaching themselves to the pipe occasionally create significant load and corrosion problems also. The possibility of human error can present an additional hazard. Dredging operations, ship anchors, and barge spuds are a constant threat to buried pipe, especially in heavily traveled waterways. Any repair work undertaken is extremely costly because of both the nature of the work itself and any needed interruption of maritime and pipeline traffic.

Figure 5.1. Pipeline suspension bridge across a river

Figure 5.2. Pipeline under a bridge designed for vehicular traffic

AERIAL CROSSINGS

An aerial river crossing is a river crossing technique in which the pipeline is either suspended by cables over the waterway or attached to the girders of a bridge designed to carry vehicular traffic.

Free-span pipeline river crossings are similar to suspension bridges designed for vehicular traffic. In place of the roadway, however, one or more pipelines supported by cables are placed over the waterway (fig. 5.1). This type of crossing has become increasingly rare, but it does still have limited application. For example, if a pipeline must cross a gorge with nearly vertical cliffs, free spans are useful and economical. Where river banks and bottoms are of solid rock, bridging the river with a span may prove to be less costly in the long run than having to blast a ditch through rock.

Despite their usefulness in these special situations, free spans are generally restricted to isolated areas. High maintenance costs, accidental damage due to wind and bending stresses, and vulnerability to vandalism all weigh heavily against the use of this type of crossing.

Major rivers and waterways are usually crossed by a number of bridges designed to carry vehicular traffic. Depending on the size and type of these structures, pipelines may be carried on the girders along the underside (fig. 5.2). While this affords greater protection and support than free spans, many of the same problems remain. Vibration from passing traffic may cause serious complications, and movement due to thermal expansion of both the pipeline and the supporting structure can also hasten metal fatigue and stress failure.

Figure 5.3. Aerial view of the staging area for a river crossing

CONVENTIONAL CROSSINGS

Most waterway crossings are installed by conventional, or normal, pipeline, methods; that is, the crossing is graded and ditched; the pipe is welded, tested, and pulled across the stream; and then it is tied in on each side to the cross-country line. Most of the steps preceding this type of river crossing follow essentially the same sequence as that for onshore construction, and if the crossing is large enough or if several crossings are to be constructed, the river crossing crew is essentially self-supporting and functions as a small pipeline spread.

The area on each side of the crossing must be cleared and graded not only to allow movement and storage of the excavated spoil but also to provide the necessary space for staging activities (fig. 5.3). Pipe, which is stockpiled in this area before preparation of the crossing section, is usually of higher tensile strength and heavier wall thickness than normal because of the potentially greater stresses and corrosion hazards. The heavier wall thickness also serves to offset the buoyancy of the pipe in water or in the soil-water mixture of the backfilled trench.

Sometimes additional weight must be added before installation to combat the buoyancy problem and prevent the section from floating out of position. Concrete weights that are either set on or bolted on the pipe are commonly employed for this purpose, though a continuous coating of concrete can also be used. Specifications generally call for a negative buoyancy 1.25 times the specific gravity of water. In large-diameter pipe, weights of 10,000 pounds each spaced on 8-foot centers or a continuous concrete coating 9 inches thick may be required (fig. 5.4).

Figure 5.4. Pipe used for river crossings has both heavy wall thickness and a thick layer of concrete.

82

Figure 5.5. Draglines are widely used in river ditching operations.

Handling joints of pipe that may weigh as much as 20 tons or more requires a stable and solid footing for a staging area. If stable, solid ground is not available at the crossing site, then sand or riprap material can be brought in to remedy the problem. Otherwise the pipe must be moved to the river's edge by side booms from the nearest area with a good surface or floated and pushed through the wet area. The terrain adjacent to the proposed crossing is very important and can contribute greatly to the problems and costs associated with water crossings.

Once the staging yard has been selected and prepared, the pipe is hauled in, lined up, and welded into

Figure 5.6. Drill barges are used when explosives are needed to excavate the pipeline trench.

one long section or several sections if the crossing is very long. If the joints have been coated at the mill, the welds must be field coated, and if the pipe is to be coated in the field, it must be done now. Then, weights are attached, or the continuous concrete coating is applied. Next, any bends or fittings required to accommodate the grade of the crossing are welded in, and the pipe is then hydrostatically tested before it is pulled across the river. Testing at this point is an extra safety precaution designed to minimize costly repairs needed to repair a leak once the pipe is in place.

While the pipe section is being readied on the bank, the ditch crew excavates the trench across the water crossing. The type of equipment and the best trenching method vary greatly with the particular characteristics of each river crossing. If the soil is stable and the river current is not too swift, the ditch can be excavated with a backhoe or dragline. If the water is very deep, then the machines are mounted on floating platforms, or barges, and the trench is excavated from those (fig. 5.5). Rock requires drilling and shooting, just as it does on land (fig. 5.6). If there is sandy unstable soil and the riverbed is wide and flat and the water relatively

shallow, a combination of well points and sand bars may be used. In this procedure, the channel is diverted to one side of the riverbed by the construction of temporary sand bars and cofferdams with draglines and bull-dozers (fig. 5.7). After the wet soil has been stabilized with well points, draglines are used to excavate the trench. When the operation reaches the diverted channel, the process is reversed, and the ditch is completed.

Figure 5.7. The river channel is diverted to facilitate installation.

84

If the channel cannot be diverted or when the volume of material to be removed is large, a suction dredge is used. Large amounts of soil are rapidly forced by a suction pump into a discharge pipe for deposit on the adjacent bank. A cutterhead, an egg beater type of attachment that generates a churning action, can be used on the dredge when hard soil is present (fig. 5.8).

Cover requirements vary with the permit issued by the controlling authority for the particular water crossing. Generally, however, they range from 5 to 15 feet. The required ditch width is usually the same as for land operations unless the soil is unstable. In unstable soil, the slope of the ditch needed to maintain cover requirements may be as great as 5 to 1.

The depth of the ditch and its alignment between the banks is closely monitored, and frequently a laser sighting device is used to keep the crossing on course (fig. 5.9). Sounding devices and divers are also employed to measure the dimensions of the ditch. When it is determined

Figure 5.9. Lasers are routinely used to guide dredging operations.

that the ditch has been excavated to the proper depth and position, installation operations begin.

Several different methods can be used to install the pipe across the river (fig. 5.10). On smaller crossings and on lines with small-diameter pipe, the section can be picked up by a number of side booms and walked across the river and lowered into place. If the riverbed cannot support the tractors, then the machines stop at the water's edge, and the pipe is floated and pulled the rest of the way across the river. Another method calls for pulling the entire section across the waterway. Side booms or rollers placed under the pipe position it for the pull across the channel. Flotation devices are attached to the section to provide lift to the pipe as it is pulled across the water (fig. 5.11). The amount of lift needed must be figured accurately. Too much lift can result in losing control of the pipe string if there are strong currents in the river. Too little lift, and the pipe may become difficult to pull.

Figure 5.8. The cutterhead, or egg beater, on the dredging unit

A

B

Depending on the size of the crossing and of the line, either bulldozers or a heavy-duty winch is used to provide the pulling power needed to move the section across the river. A plug, or hairpin, is welded to the front of the pipe section, and cables are attached and strung across the river to the winch. As the pipe is pulled slowly across the waterway, other cables may be attached to hold it in place in the ditch line. A tremendous amount of coordination and timing is needed to guide the pipe into the correct position, and often the crew gets only one chance. Once the line is in position,

Figure 5.10. Two methods of installing pipe across a river channel: *A*, walking; *B*, pulling.

Figure 5.11. Flotation devices are attached to the pipe before it is walked or pulled across a river.

quick-release mechanisms attached to the flotation devices are activated, and the pipe sinks into position in the ditch.

When the pipe is in place, the same equipment used for excavation begins to backfill the line. Because the pipe can be forced out of the trench with too great a volume of loose spoil, backfilling is a slow and precise operation. In rocky areas, special backfill may be required to protect the pipe. Generally, however, the excavated spoil is returned to the trench and placed over the pipe.

Once the pipe has been backfilled properly, the river section is again hydrostatically tested to ensure the soundness of the crossing. After this testing process, the river section is tied in to the adjoining cross-country sections.

COMMON PROBLEMS WITH WATER CROSSINGS

The hazards and complications inherent in laying an underwater pipeline are compounded by a number of variables that contractors do not face in land construction. High water, for example, can cause delay even if the actual weather conditions are satisfactory. Also, the extensive navigation common on most major waterways adversely affects construction schedules. Since prolonged disruption of traffic cannot normally be tolerated, the construction schedule of the entire project can be delayed while necessary permits are obtained from both environmental and maritime agencies.

Once the pipeline is in place, ongoing maintenance is required for the line's safety and security. This maintenance involves checking for coating and corrosion problems as well as establishing adequate safeguards against mechanical damage from ship anchors and barge spuds. In addition, river bank stability must be checked and reinforced where necessary, especially if flooding has recently occurred. Stabilization is usually undertaken after the original contours have been restored to the bank. Materials designed to hold the soil in place, collectively referred to as riprap, are arranged in a way that will prevent erosion and shifting of the soil (fig. 5.12).

Figure 5.12. Riprap prevents bank erosion.

DIRECTIONALLY CONTROLLED HORIZONTAL DRILLING

Problems that eventually develop with river crossings can usually be attributed to the relatively vulnerable location of the pipe once the crossing is completed. Efforts by engineers to overcome this problem and thus reduce the potential for damage have produced a number of effective solutions. Among these is a technique that calls for burial of the pipe under the riverbed, far below the normal depth of conventional crossings. This technique is termed *directional drilling*.

While it is not feasible in every situation, where conditions are favorable directional drilling has proven to be a reliable and cost-effective procedure.

The striking feature of directionally drilled river crossings is the burial of the pipeline in an inverted arc beneath the riverbed at depths of up to 40 feet (fig. 5.13) With this level of cover, the pipe is considered to be well out of harm's way. Theoretically at least, it should not be subject to the same degree of corrosion, scouring, and other maintenance problems that might normally be expected to plague pipelines buried at shallower depths.

Figure 5.13. Horizontally drilled crossing

HORIZONTALLY
DRILLED
PIPELINE

Before the crossing is begun, drilling equipment, auxiliary machinery, and support structures are set up on one side of the river. The drilling rig itself is mounted on a trailer to facilitate transport and access to the job site. Depending on the design of the crossing, the rig can be elevated to allow it to bore at the best angle for putting through the pipe joints (fig. 5.14). To one side of the rig are mud pumps and holding tanks. During drilling, mud is pumped into the hole both to lubricate the drill bit and to draw off the cuttings, or spoil.

The initial step in the process is to complete a hole from the rig side to the opposite bank. This pilot hole is dug by a relatively small rotating bit (2 inches or so in diameter) powered by a hydraulic motor. As this drill assembly moves forward, joints of small-diameter pipe, making up what is called the pilot string, are joined behind it. The pilot string is made up one joint at a time and is threaded onto the preceding joint by a set of tongs on the rig. The rotating action of the rig connects the new joint and drives the others deeper into the hole. The pilot hole is completed when the head end surfaces, or punches out, on the opposite bank. The main function of the pilot string is to establish a pathway for the pipeline.

A computerized guidance system is used to maintain the correct course for the pilot string and for the carrier pipe later on. The computer processes information fed to it by instruments such as a pendulant, which determines inclination, and a gyroprobe, which is sensitive to drift and bearing. Sonar can also be used to plot azimuth, the bearing of the pipe in angular degrees from a north-south axis.

Guidance systems can now give an impressive degree of accuracy consistently. Even on crossings greater than one-half mile, the pilot string often punches out within inches of the marker stake.

The pilot string is only temporary, and it is removed after it has established the route. This removal begins during the second stage of drilling. The drill bit and motor are removed, and an adaptor is used to attach pipe of slightly larger diameter, known as washpipe, to the pilot string. By reversing the direction of rotation on the pilot string, the rig can pull the washpipe back under the river. As it is pulled through, joints of washpipe

Figure 5.14. Head-on view of a drilling rig

Figure 5.15. A pullback assembly is composed of the fly cutter, the barrel reamer, and the swivel.

are added on the pipe side of the crossing. At the same time, the pilot string is disassembled on the rig side.

The washpipe makes up what is referred to as the *work string*. This pipe remains in place until the actual pipeline is made up and ready to be pulled back across the river. Although the washpipe used can vary in diameter and wall thickness, the key consideration is to select pipe with sufficient tensile strength to resist coming apart during the pullback procedure.

A special three-part connection joins the washpipe and the pipeline during the pullback. A swivel assembly is attached to the pipeline by means of a hairpin, or U-shaped tongue, welded to a cap over the pipe end. A cylindrical barrel reamer, mounted immediately in front of the swivel, is fitted on both ends with hollow cutting teeth. Drilling mud pumped through the work string into the barrel reamer exits at high pressure through these teeth, thereby lubricating the hole and carrying off the spoil. The barrel reamer normally has a diameter slightly larger than the pipeline and serves both to open the hole and to keep the pull on course.

A cutterhead, or fly cutter, is the lead component in the assembly and does the actual boring of the hole. The cutterhead is a circular steel band ringed with conical cutting teeth. Three supporting bars, or spokes, on its inside diameter are also studded with double rows of teeth. Drilling mud is pumped through these teeth also, and the clearing of the hole proceeds in a manner similar to slick boring in road crossings (fig. 5.15).

During the actual pullback, the rig does the heavy work. By generating pull and rotational force on the work string, it brings both the pipe and the spoil back across. It is important to note that although the fly cutter and barrel reamer are rotated at a fairly high rate, the pipeline itself does not rotate. The hole is made to provide adequate clearance for the pipe being pulled through, and the diameter is usually 10–12 inches greater than the diameter of the pipeline.

The potential applications of directional drilling are no longer limited solely to river crossings. A beach or shore approach, where the pipeline is brought from an offshore well onto land, is one of the more promising possibilities. In the Arctic, some form of directional drilling may also prove to be a workable solution to the problem of working around large formations of pack ice.

Swamp and Marsh Construction

Construction through swamps and marshes is a hybrid type of pipelining. Elements common to offshore work, river crossings, and even cross-country jobs are all combined to some extent. Although swamp and marsh construction both involve working in wet, unstable soil conditions, it is important to distinguish differences between the two.

Swamp construction implies the presence of trees in and along the right-of-way. Typically these are large cypress trees with extensive root systems. Because of their size, clearing is done with backhoes and dynamite. The stumps and spoil are then removed and placed off to the side of the right-of-way. Marshes, in contrast, are watery areas where grasses and other water plants are the predominant vegetation. Extensive clearing is usually unnecessary, since the machinery that makes the ditch can accomplish both tasks.

The water is always shallow in either type of environment. In swamps especially, stretches of flooded right-of-way may be interspersed with patches of water-logged, unstable soil or quicksand.

Although water is normally considered to be an obstacle in other areas of pipelining, in swamp and marsh construction it is essential to the successful completion of the project. Rarely, if ever, is there any advantage to draining a swamp or marsh before beginning work. In fact, drainage ditches that happen to intersect the right-of-way are often plugged to keep as much water as possible in the pipeline ditch. Without adequate water depth it becomes impossible to float the pipelaying machinery, the support equipment, or the pipe itself. Thus, when the pipeline must pass through an inundated area, too much water is actually preferable to too little. The contractor could be left high and dry if, for example, a shift in wind direction should occur and blow the ditch water out to sea. The result would be a right-of-way still too wet to attempt land construction but not wet enough for swamp and marsh techniques.

Figure 5.16. If possible, a suitable area of dry ground is used as a staging area for swamp construction.

CONSTRUCTION METHODS

The condition of the right-of-way and the owner company's contract specifications will determine the best method of laying the pipeline. Generally, marsh and swamp construction is a push operation; that is, the pipe is brought in to the job site, welded together, and then pushed down a water-filled ditch (fig. 5.16). More specifically, pipelaying operations are accomplished in one of three ways: pushing pipe from a land ramp, pushing pipe from a floating lay barge, or laying pipe in a canal.

In order for the ditching equipment to work effectively, a firm, stable surface is necessary. Mats made of large timbers are bolted together and laid down to form a flat, wide surface (fig. 5.17). These enable the ditching equipment to walk across the swamp as the ditching progresses. Although the mats are cumbersome, they are relatively portable when handled by backhoes and similar heavy equipment. During excavation the ditch walls are sloped to the point where they will stand without collapsing. In marsh ditching, so-called marsh buggies are used. Pontoon floats, made integral with the buggy, permit it to float on lakes or streams, and mats are not required.

Crossing other pipelines that intersect the right-of-way demands extraordinary caution and special equipment to handle the pipe. The owners of these lines specify beforehand the amount of vertical clearance to be maintained between the two lines. Whether crossing above or below the line, 2 feet of clearance is standard. If the pipe is to go underneath the foreign line, the ditch is gradually angled deeper, starting

Figure 5.17. Large mats provide a stable base from which the dragline works.

several hundred feet from the point of intersection. It is important for the slope of the ditch to taper gradually, so that the slope will allow the pipe to conform, or lay, to the contour of the ditch without actually bending the pipe. While the majority of contractors cross a foreign line by passing underneath it, occasionally an owner may stipulate that no digging is to be done beneath the line already in place. Usually this is the case with large-diameter pipelines, since it is believed that the chances of damage are increased if any excavation is attempted. The pipeline is then laid above the other line, with an adequate amount of clearance between them. The choice will generally be set forth in the owner company's contract specifications.

Pushing pipe from a land ramp shares some of the same techniques found in conventional river crossings. If a sizable area of accessible, dry ground is near the swamp or marsh, a central staging area where the joints of weighted pipe are welded into the line can be established. A concrete coating is applied to the joints at the mill before they are delivered to the site.

Figure 5.18. Flotation devices attached to the pipe before it leaves the barge facilitate the push down the ditch.

Sufficient floats are attached to the pipe as the pipe is pushed into the ditch, so that the pipe will float (fig. 5.18). Once the pipe has reached a designated point, the flotation devices are released, and the pipe is allowed to sink.

Pushing pipe from a floating lay barge employs the same techniques as pushing from a land ramp. The lay barge consists of several barges that are connected end to end so as to make a working surface 300 to 400 feet long—the same length as the corresponding land ramp. The lay barge is able to anchor in slips dug at crossing waterways and to then push pipe into the connecting marsh or swamp ditch (fig. 5.19). Pipe and other supplies are brought to the lay barge on supply barges.

In the canal lay method, by contrast, the forward motion of the barge sends the pipe down the ramp to the water. Because the lay barge moves along the entire length of the right-of-way, the ditch must be deep enough and wide enough to float the barge. Every time a joint of pipe is completed, the barge is moved ahead roughly 40 feet. This technique is also referred to as marine lay.

This coating, which affords the additional weight and protection that submerged pipe requires, is applied over such corrosion-resistant coatings as coal tar enamel, heavy mastic, or a thin-film epoxy. After the pipe is welded, the field joints are coated with a material compatible with the mill coating on the pipe. Then, a concrete coating is usually applied by placing a steel sleeve over the joint area and filling the void with concrete.

Anchoring devices called spuds hold the barge steady and keep it

Figure 5.19. Lay barge

aligned properly. Spuds are long steel structural members that extend vertically from two corners of the barge. When the barge is to be moved, they are hoisted by powered winches that lift the spuds. When the barge is to be secured again, the lift mechanism is released, and each spud's own weight again buries it in the bottom.

No flotation devices are attached to the pipe coming off the lay barge—the pipe is placed directly on the ditch bottom. Pipe already in place gives the barge a solid foundation against which it can winch itself forward. A tugboat can also be used to move the barge.

When pushing pipe from a floating lay barge or while laying in canal, a large crane unloads the pipe from a pipe barge anchored alongside the lay vessel (fig. 5.20). A side boom on the lay barge aligns the joint and starts it moving through the separate work stations set up along the entire length of the barge (fig. 5.21). The stringer bead is made in the first of several welding stations. Once the joint is welded onto the pipe string, it moves to X-ray inspection and finally to coating and sealing of the weld. X rays are developed and examined immediately to detect weld defects before the pipe is submerged.

In canal lay, barge-mounted clamshell excavators (5 to 6 cubic yards in size) are used to clear the timber as required and to excavate a canal sufficient in depth (7 to 8 feet) and in width (40 feet) to float the clamshell barge. Barge-mounted dragline excavators (5 to 6 cubic yards in size) are then used to excavate the ditch for the pipeline within the limits of the canal. The lay barge is then positioned in the canal. Pipe is brought to the front end of this work platform by smaller barges that shuttle back and forth carrying pipe and supplies. The pipe joints are then fed through the various work stations on the main barge. As the initial joints are completed, they are pushed off the end of the barge and back down the ditch to a specific point. The barge then begins to move in the canal, laying pipe away from that point. The sunken pipe is again used to push off and propel the barge forward.

Figure 5.20. After being brought to the lay barge by a supply barge, joints are unloaded by crane and placed in a holding area.

Figure 5.21. A side boom moves the joint into position for alignment.

BACKFILLING

The same equipment used for ditching is used for backfilling. The amount of spoil available to refill the ditch, however, is invariably less than that which was originally taken out. This loss is due to the compressing effect of the backfill as it is piled on the spongy soil next to the right-of-way. Bringing dirt in to make up for this shortfall is impractical, and so a small depression usually remains visible on top of the ditch along the length of the pipeline.

HYDROSTATIC TESTING

The pipeline is tested in as few test sections as possible because tie-in welds are extremely expensive in swamp and marsh construction. Standard hydrostatic test procedures are employed.

Offshore Pipelines

If the demand for energy for the rest of this century and beyond is to be met, technology for locating and tapping natural resources in remote locations must be developed. Among the areas that will remain at the center of resource recovery efforts are the world's oceans. Remarkable strides toward more fully exploiting these resources have been made in offshore oil and gas exploration technology. The pace of offshore technology is so great that oil and natural gas from marine sources is soon expected to account for at least half of total world demand. Regardless of offshore locations, the hydrocarbons produced must somehow be brought ashore.

Bringing offshore production ashore is the job of deepwater pipelines.

In contrast to swamp and marsh construction, offshore pipelining involves working in water depths of 100 feet or more. Up until about 1967, the practical limit for subsea pipelines was considered to be approximately 175 feet. Now pipelines are routinely being laid at depths of up to 600 feet, and experts predict that future depth capabilities will be in the range of 3,000 to 6,000 feet. The ultimate pipeline depth appears to be contingent mainly on the economic feasibility of the project. Certainly future advances will not be impeded by lack of engineering ingenuity.

Weather conditions affect all types of pipeline construction to a greater or lesser degree. It is safe to assume, however, that no other type of pipe laying is more vulnerable to weather-induced setbacks than offshore construction. Installation of subsea lines follows the same basic sequence used in land construction. Line-up, welding, X-ray inspection, and coating of the welds must all be completed on the lay vessel or platform before the pipe string is submerged. The high cost of laying pipe in this environment, however, allows virtually no latitude for delay, regardless of the cause. Thus, the overriding consideration in offshore work is to lay the pipe as rapidly as possible while providing the safest possible working conditions for all operating personnel and equipment.

LOGISTICS

Crucial to the success of an offshore pipeline is precise coordination of logistical support. Offshore pipe-laying vessels resemble floating cities.

As such, they are designed to operate for extended periods in open seas. Limited storage space aboard the vessels and the capability to lay as much as a mile of pipe a day means that resupply needs are virtually constant. Thus, a single lay vessel often requires three or four supply ships as well as auxiliary vessels such as tug boats and survey ships.

LAY BARGES

Of the several vessel configurations currently in use offshore, the lay barge is by far the most effective over a wide range of conditions. It owes its versatility to economical operation, simplicity of design, and a relatively large and stable work area for storage and movement of materials (fig. 5.22). In addition it can be used in many different water depths, an advantage when pipe is being laid from relatively shallow water near the shoreline to a production rig in deeper water farther out.

The main deck of the lay barge can be thought of as an offshore pipeline spread. While not every operation in land construction has an exact equivalent offshore, most of the principles governing pipe handling and makeup are the same. A more extensive array of pipe-handling equipment is needed offshore, however, to deal both with problems of ship-to-ship transfer of materials and with fickle weather conditions.

The pipe-handling equipment on lay barges consists primarily of heavy-duty cranes for hoisting the concrete-coated pipe joints. The joint is carried by conveyor from the storage rack to the line-up station, where the line-up shoes align the end of the pipe with the joint ahead of it in the string. An

internal clamp identical to those used in land construction maintains this alignment while the root bead is made at the first welding station.

As in other types of pipelining, productivity increases offshore depend heavily on improvements in welding methods and techniques. Pipe-joining methods for offshore lines have undergone significant changes, and continuing research offers the promise of even greater gains as new methods become available.

The mainstays of offshore pipe joining are automatic welding systems, so called even though a certain

Figure 5.22. Center-slot pipe-laying barge

amount of manual adjustment is usually necessary. A definite trend in offshore welding is the reduction of manpower requirements. At the same time, weld quality and productivity rates are being maintained or are increasing. Theoretically, fewer work stations and automatic welding systems should increase the rate at which the pipe is laid.

Generally speaking, automatic welding systems fall into one of three classifications: gas metal-arc welding (GMAW); gas tungsten-arc welding (GTAW); and flash-butt, or simply, flash welding (FW), which is the most recent innovation. Except for minor variations, the gas arc systems essentially follow similar land operational sequences. Because each uses inert gas to shield the weld as it is made, these processes are often collectively referred to as shielded metal-arc welding (SMAW). Shielded arc welding is currently the most widely used method for offshore work. Its popularity is due in large part to a relatively uncomplicated technique that consistently produces high-quality welds regardless of the type or grade of pipe being welded.

The ultimate goal of pipe-joining research is to develop a process capable of joining the pipe ends together at a single station. Not only would this decrease labor and capital expenditures; it would also open the way for vertical installation of large-diameter pipe in very deep water (over 2,000 feet). Currently the most promising techniques for achieving single-pass pipe joints is flash welding.

In flash welding, low voltage is applied to each pipe joint while the ends are in light contact. This contact is sufficient to produce a rapid arcing, called flashing, between the pipe ends. After the pipe ends have been adequately heated, the current is abruptly increased, and the pipe joints are brought together rapidly and forcefully. After a short time, the current is reduced. A machine automatically clears excess flash material from inside the pipe, and the weld is completed.

Further experimentation with alternative pipe-joining methods is continuing. Among the proposals under study are those that would join pipe using exotic mediums such as electron beams, lasers, and plasma arcs. The success of these methods will depend both on scientific proof of their feasibility and, of equal importance, on their widespread acceptance by the industry.

The massive financial commitment required to operate a marine pipeline makes strict weld inspection imperative. Any repairs needed after the line is in place will compound the financial burden as well as disrupt the line's operation. Thus, radiographic inspection is mandatory for all field girth welds. Standard X-ray inspection methods using an external radiation source are acceptable though somewhat slow. Internal X-ray crawlers are usually preferred because they allow extremely short exposure times. Film is developed immediately and evaluated for evidence of weld defects.

Before the pipe is submerged, a protective coating is applied over the weld. Offshore field coating is done in a manner similar to that used in some swamp and marsh work.

Except for the ends, the pipe joints used offshore are usually concrete-coated at the mill before delivery by the supply vessel. The thickness of the concrete needed depends on its density, the size of the pipe, and the presence and strength of any deep currents in the area. Where currents are strong, pipe stability becomes an important consideration. Subsea mud, silt, and sand tend to behave like dense, viscous fluids in the presence of strong currents. Movements of these materials can undermine or shift the pipeline. If such movements leave an unsupported pipe span, spalling and mechanical damage can result. Concrete coating also serves to protect the pipe during the actual laying operation. During laying, pipe is subjected to the maximum stress that it will ever receive. The coating must be able to absorb these stresses without compromising the integrity of the pipe.

The deck area of a lay barge is used to assemble the pipe joints into a continuous string in what is referred to as the stovepipe method. Individual work stations are spaced along a gently sloping production ramp through which each joint passes. Differences in the number of work stations on lay barges is primarily a reflection of the type of welding process used. The welding process also dictates the length of the production ramp and, to a certain extent, the size of the barge itself.

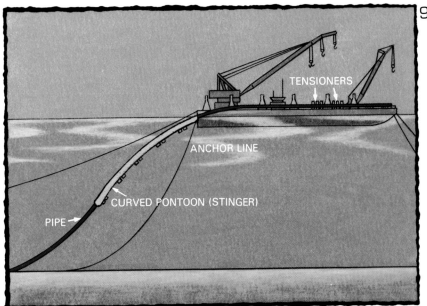

The most critical step in the pipe-laying operation begins when the pipe enters the water. If this procedure is mishandled, either the coating or the pipe is likely to be damaged. To minimize stress on the pipe joint as it leaves the barge, an extension of the production ramp, called a stinger, supports the pipe on its way to the ocean floor (fig. 5.23). The stinger is hinged on the stern of the barge to permit adjustments in the angle of pipe launch according to the water depth. Regardless of the depth, the pipe is launched in an S-curve configuration, with an overbend closest to the barge and a sag bend at the sea bottom (fig. 5.24).

Figure 5.23. Stinger and lay barge with stinger system

Figure 5.24. S-curve method

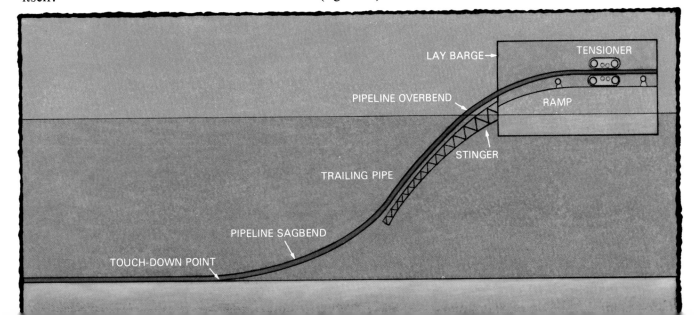

Stinger length varies and may extend the entire distance to the sea bottom. More often, though, stingers are built to be just long enough so that the lead end of the pipe rests on the ocean floor while the other end is still on the stinger. The temporarily unsupported span of pipe between the stinger and the seabed is called the sag bend, and its maximum permissible length is computed beforehand. This information is used to determine the size of the stinger and the maximum water depth in which a given stinger can be used.

The main objective in each design is to make best possible use of the entire stinger length. This is readily accomplished by increasing the angle at which the pipe exits the barge. Short curved stingers, for example, and those made up of articulated segments are able to increase the lay angle while still retaining the capabilities of longer stingers.

Innovations in stinger design account only partially for the rapid improvements in deepwater pipelining. The capabilities of auxiliary equipment such as pipe tensioners have also had considerable impact. Tensioners act essentially as braking devices by applying external pressure on the pipe string to control its descent rate. They also support the entire submerged weight of the pipe as it approaches the bottom.

Stronger tensioners have been developed to reduce the degree of sag bend and thus limit the stress on the free span of pipe between the bottom and the stinger. Increased tensioner capacity also allows the pipe to be placed in the water at an angle closely approaching 90 degrees. Shorter stingers can then be adapted for use in deeper water. Used in tandem, modern tensioners can increase holding capacity to nearly 500,000 pounds—enough to safely support large-diameter pipe in very deep water.

By means of a sophisticated monitoring system, most offshore barges now use dynamic tensioning to control pipe release from the stinger. Wave motions that produce sudden movements of the barge—either horizontally or vertically—are immediately compensated for either by paying out the pipe string or by hauling it back. A constant level of tension is thereby maintained on the pipe to prevent slack sections from developing. Thus, the tensioners help safeguard the pipe from wall collapse, or buckling, to which it is particularly vulnerable during its release from the stinger.

Many of the advantages of laying pipe with a shorter stinger are enhanced when a steeply inclined ramp is used instead. Ideally, the pipe should be laid vertically with only a single work station needed for joining the pipe. Until such a system is adequately field tested, however, the inclined ramp method offers an acceptable alternative to true vertical pipe laying.

An inclined ramp attached to the barge stern is able to rotate as much as 60 degrees in the vertical plane. This arrangement eliminates the stinger, and the pipe enters the water in a J curve without overbend. Tensioners bear the weight of the freely suspended pipe as it is lowered to the bottom.

ANCHORING SYSTEMS

In offshore construction, it is the motion of the lay vessel and not of the pipe itself that puts the pipe into the water. The apparent movement of the pipe along the production ramp is therefore somewhat illusory. The lay barge itself is not self-propelled, and so some other source of motive power is needed. The solution is an elaborate anchoring system that is capable of holding the vessel on station as well as moving it forward along a specific route. Depending on the size of the barge and the mooring system used, a lay vessel may require as many as a dozen or more anchors. Because of the dual purpose these anchors serve, they bear little resemblance to those commonly seen in lighter marine work.

Lay-barge anchors can weigh in excess of 20 tons. The barge is tethered by at least one anchor in each corner, as well as by breast anchors that secure it amidships on both the starboard and port sides. A second important component in this system is the anchor wire and winches that pay out and reel in the anchors. The depth of the water determines the length of line needed, which in turn dictates the winch size. Even with a heavy anchoring system, some lay vessels still require tugboat assistance to keep them on station.

When the barge is working, tension on the anchor lines holds it in position. As each joint is completed and is ready to be submerged the barge pulls itself forward by slacking off on the stern anchors and picking up on the bow lines. The sequence is actually more complicated, especially in deeper water. Miles of cable may be involved, much of it between 2 and 3 inches in diameter. For that reason, mooring systems control is increasingly being turned over to computers. High-speed data analysis allows for more rapid adjustments of winch tension and barge position.

SEMISUBMERSIBLES

A major disadvantage of all types of surface pipe-laying vessels is their high susceptibility to adverse environmental conditions, particularly wave action. In the areas where drilling is now being done, roll and heave motions can shut down operations for extended periods at considerable financial loss to the pipeline owner and contractor.

Borrowing from technology developed for offshore drilling rigs, semisubmersible lay vessels are now used where conditions are expected to be continuously rough (fig. 5.25). The submerged pontoon portion of the hull and the elevated work area gives the vessel superior stability even under the most severe conditions. This type of hull is very effective when used in conjunction with dynamic positioning systems.

Figure 5.25. Semisubmersible pipe-laying vessel

99

A potential problem that may develop with the use of semisubmersibles concerns resupply complications. For even though the lay vessel remains relatively steady in high seas, it is unlikely that supply ships can operate if conditions are especially bad. One solution has been to increase the storage capacity for essential materials on board the lay vessel to ensure some degree of self-sufficiency should the situation demand it.

REEL BARGES

Heavy wall thickness and high tensile strength are among the most important characteristics for line pipe used in offshore work. These characteristics produce a rigidity and strength essential to the pipe's withstanding stresses generated during the pipe-laying process and later throughout the service life of the line. The actual strength and flexibility of this pipe is clearly demonstrated by its ability to be wound onto a reel for use on a reel barge.

Makeup of the pipe string for a reel barge is done onshore at a central assembly area. Welding, the application of thin-coat epoxy coatings, and testing are all handled routinely, and then the pipe is coiled onto an immense reel (fig. 5.26). Although this procedure does cause some wall thinning, its effect is considered insignificant. If anything, the pipe may be strengthened by this cold working.

A lay barge specially outfitted for this type of pipe laying then transports the reel to the lay site. It is a relatively simple procedure to pay out the pipe at a steady rate onto the ocean floor. In fact, compared to more conventional methods, little time is spent doing the actual laying. The main time expenditure involves winding the pipe onto the reel and bringing it to the job site.

Figure 5.26. Reel barge and reel

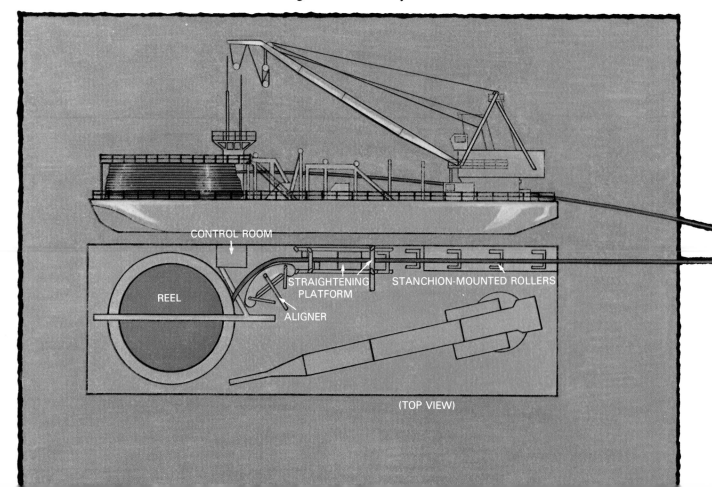

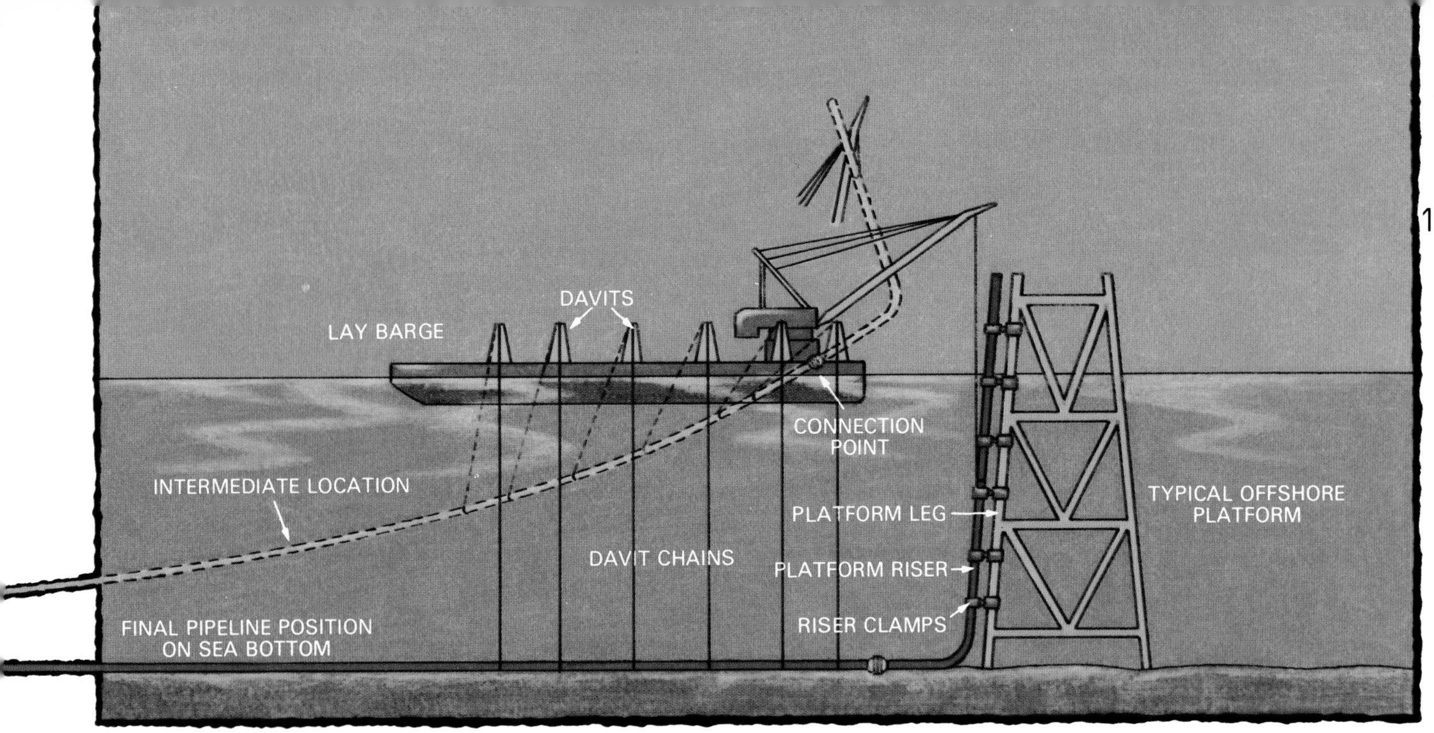

Figure 5.27. Platform riser installation

The reel method is expected to account for a much higher percentage of new offshore pipeline mileage in the future. Continued improvements in pipe metallurgy and barge design should open the way for use of this method for laying large-diameter pipe of 30 inches or more.

BOTTOM-PULL METHOD

In an attempt to lessen the risk of downtime due to unfavorable weather, some pipeline companies have shown a renewed interest in the bottom-pull method in which the pipe string remains below the surface as it is towed to its final location. While this technique has long been popular for pipe laying at shallow depths, its adaptation for deeper water pipelines is fairly recent.

Before the actual laying operation is begun, the joints of line pipe are welded together into a single long section, or several shorter sections, at an onshore location. The makeup area requires direct access to the shoreline, usually through a beach approach.

When the pipe section is completed, a pulling head is attached to the lead joint in the string. Flotation devices or pontoons may also be spaced along the length of the pipeline. These provide a specific degree of lift for each joint depending on the total weight of the pipe string and the pulling capability of the tow vessel. The pipe thus may ride directly on the bottom or at some intermediate depth. When it is decided that the pipe will be towed along the bottom, trenching equipment may be brought in to smooth out, or *presweep,* the route.

The potentially weakest part of the entire pull system is the link between the tow vessel and the pipe. Multiple chain or cable connections are used to ensure that the massive weight of the pipe string does not cause it to pull apart or separate from the tow vessel.

Regardless of the method of installation, the offshore pipeline is eventually joined to a subsea riser on the platform. Risers are vertical sections of pipe that connect the pipeline on the sea bottom to the production platform on the surface (fig. 5.27). The riser is an integral part of the pipeline and is clamped directly to a leg or brace on the platform. Some platforms also have a shoe, or guide tube, welded onto them when they are first set up.

As more drilling is done in deeper waters, the length of the subsea riser increases proportionately, and this creates serious handling problems. To circumvent these difficulties, several options are available for riser installation. In the J-tube method, the pipe is lifted off the ocean floor when it reaches the platform and then is fed up to the surface through a guide tube. The pipe assumes a J configuration, from which the technique derives its name.

Another option is the reverse J-tube method. In this variation, the pipe is welded together on the platform itself and then fed down through a guide tube to the seafloor. A pulling head on the lead joint is attached by a cable to the tow barge. The barge pulls the pipe away from the platform, one joint at a time. When the pipe makes land, it will connect with an onshore pipeline. It may also be joined to a previously laid marine line at sea.

Larger-diameter pipelines, deeper waters, and longer pipe strings generally are not compatible with the preceding methods. A major objection is that any procedures involving additional handling of the pipe pose unacceptable risks. Research has therefore focused on ways to join the pipe and riser on the seafloor without additional lifting or manipulation.

One approach to this problem is to install the riser on the platform separate from the pipeline. The basic idea is to substitute the riser for the guide tube. Welding is then done underwater in a dry atmosphere habitat by welder-divers specially trained in saturation diving techniques. Saturation diving involves breathing a mixture of nitrogen and oxygen for extended periods underwater. The ac-tual welding takes place in a pressurized chamber that is lowered over the side of the platform and locked onto both pipe and riser. X-ray inspection, testing, and even coating can all be carried out in this pressurized environment.

REMOTE CONNECTION

A more recent pipe-joining innovation that has been successfully used in conjunction with bottom towing is remote connection. As the name implies, the connection process is directed from somewhere other than at the immediate site—in this case, from a control panel on the platform deck. A sled, or connector, fitted to the front of the pipe string is designed to link up with a receiver connection located on the end of the riser.

A connection may be completed in one of two ways. The first of these, termed a direct pull-in, involves steering the pipe with its connector sled straight onto the matching receiver at the base of the platform. The second option is called the deflect-to-connect technique, in which the pipe is pulled to a target area in line with the platform but to one side of it. The final connection is made by winching or otherwise deflecting the pipe laterally until it mates with the riser connection. With both techniques, the connection is completed either by mechanical connectors or by welder-divers.

TRENCHING

Unlike most land construction, offshore pipe laying does not always require that the pipe be buried. The determining factors tend to be the risk to the pipe at a particular water depth.

Anchor damage from fishing activity or other types of marine work is usually avoided by burying the pipe. Significant seismic activity or undersea mud slides also can endanger the pipe.

The sequence for trenching underwater lines is the opposite of that used onshore. Marine pipelines are trenched after the pipe is laid, and a suction dredge or bury barge is used. The bury barge moves itself forward with an anchoring system, as does a lay barge. Attached to the bury barge by a cable is a jet sled, which straddles the pipe (fig. 5.28). The jet sled is fitted on both sides and inward-directed nozzles to which surface hoses are attached. Massive-volumes of water under high pressure are pumped through these nozzles to remove spoil from beneath the pipe. The spoil is then pumped to one side and the line sags naturally into position in the trench.

Figure 5.28. Pipe-burying operation showing jet sled

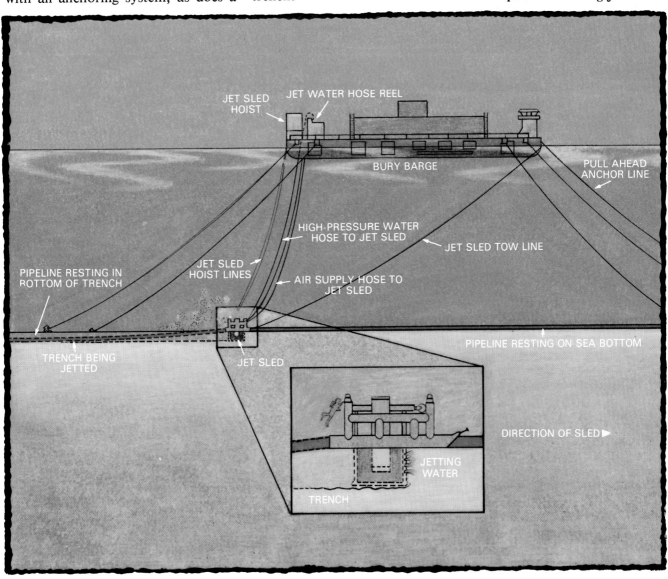

Figure 5.29. Divers are sometimes used to inspect pipelines.

INSPECTION AND TESTING

The potentially disastrous environmental and financial consequences of pipe failure ensure that strict standards are maintained during the testing and periodic inspection of an offshore pipeline. Shortly after installation, the line is pigged several times to check for buckling damage and to clean the internal surfaces. Hydrostatic testing is done in accordance with the regulations governing the location and type of pipeline. Visual inspection of the pipeline is occasionally warranted when damage is suspected or when cover depth must be verified (fig. 5.29).

Another device that has proven its value in undersea inspection is the submersible. Submersibles are essentially two-man minisubmarines. Their chief advantage is their ability to dive to greater depths and remain on the bottom longer than would be possible for divers. In addition, they can carry extensive photographic and sonar equipment. Submersibles can also be used to survey particular stretches of a proposed pipeline route to confirm data gathered earlier with electronic survey instruments.

Arctic Construction

Like the needle of an immense compass, the search for energy sources is shifting decidedly northward. While producing areas in the deserts of the Middle East and other, more temperate climates will continue to be important, discoveries in Alaska, northern Canada, and the Arctic show the potential for massive production of both natural gas and oil (fig. 5.30). This interest in the north has grown steadily and is the result of a complex combination of a number of political, economic, technological, and geological factors.

Already this sand-to-ice transition has had far-reaching effects on pipeline design and research, and the effects of this trend will continue to be felt as the pace of Arctic exploration and construction intensifies. The terrain and climate of the Arctic present unique engineering problems for pipeline contractors (fig. 5.31). Arctic construction is further complicated by the mandate to maintain high environmental protection standards while working in one of the most forbidding, yet fragile, ecosystems on earth. Strictly speaking, Arctic construction includes only that work done above the Arctic Circle, or 60° north latitude. Activity in this area centers on oil fields in and around Prudhoe Bay.

Figure 5.31. Harsh conditions punish workers and their equipment.

Figure 5.30. The Valdez terminal typifies the potential of the Arctic region.

The pervading element in the Arctic region is ice—minicontinents of pack ice in the seas surrounding the Arctic Ocean and permafrost, or frozen earth, on land. In addition to ice, brutally low temperatures prevail throughout much of the year. During the winter especially, the ambient temperature seems to fluctuate only between cold and unbelievable cold. Adapting to the effects of such climatic extremes is one of the major engineering challenges of the push north.

Because of the innovations it spawned and because it was the first major pipeline in Alaska, the Trans-Alaska Pipeline System (TAPS) is often assumed to be representative of all cold-weather pipelining. Some techniques used for the first time in its construction eventually became standard for the industry. Others, however, were peculiar only to TAPS because of the magnitude of the project and its location. The main area of what can be considered true Arctic construction currently centers around the North Slope of Alaska, Prudhoe Bay, and the Beaufort Sea. Potential reserves there have been estimated at 170 billion barrels of oil and nearly 1,800 trillion cubic feet of natural gas. The pipelining techniques used there are very different from those used for the TAPS line.

ENVIRONMENTAL CONSIDERATIONS

Despite the fact that it remains frozen for much of the year, the tundra (fig. 5.32), the treeless plain characteristic of much of the Arctic, is highly sensitive to environmental disturbance. The cardinal rule of working on the Arctic tundra is *never* to disturb it—therefore, pipelines are constructed and laid aboveground. This concern for environmental protection cannot be overemphasized, because it pervades every aspect of construction work in the Arctic. Movement of equipment, for example, is done only on special haul roads constructed either of gravel (for thaw conditions) or ice aggregate and snow (for winter work). Areas of the pipeline that, because of design factors, do not lie close to the haul road must be identified beforehand and then completed from the ice roads before the thaw or before the gravel pads are constructed.

An additional environmental safeguard is a comprehensive oil spill contingency plan. This mandatory plan outlines in detail the contractor's procedure for dealing with leaked or spilled fluids. While the emphasis is on dealing with large-scale accidents, technically the contractor is responsible for any amount spilled, no matter how small. Plan approval must be obtained from the environmental regulatory agencies before construction is allowed to begin.

Figure 5.32. The tundra has relatively little snow.

VERTICAL SUPPORT MEMBERS

Vertical support members, or VSMs, are H-shaped and T-shaped structures fashioned from steel pipe that support the aboveground pipeline. They were first used on TAPS, where they achieved a degree of sophistication unmatched by any other project since (fig. 5.33). The uprights of the TAPS VSMs support freeze pipes, which continuously circulate a refrigerant fluid between the subsoil and the top of the pipe. The refrigerant keeps the ground beneath the pipeline frozen to prevent frost heaving, a type of soil movement that results from alternate thawing and freezing. Frost heaving generates stress on the VSMs and, by extension, on the pipe itself. Periodic aerial surveys are conducted with infrared camera equipment to detect temperature variations along the right-of-way.

The VSMs used in the Arctic are an outgrowth of design research and development originally done for TAPS (fig. 5.34). The Arctic tundra is not subject to the same degree of frost heaving, however, and thawing is usually confined to the first 18 inches of soil. For this reason, freeze pipes are unnecessary.

Figure 5.33. Vertical support members on the Trans-Alaska Pipeline System

Figure 5.34. Arctic VSMs

Foundation holes for support uprights are drilled to approximately 17 feet. Two methods are currently used to make the holes. The first of these uses a conventional auger.

Figure 5.35. Slide-and-guide pipe saddle

Although it is effective, it is also rather time-consuming. The auger must be frequently pulled out of the hole to clear it of tailings, or spoil. The alternative method uses a tricone drill bit mounted at the end of a hollow kelly bar. As the drill bit progresses down the hole, compressed air is forced through the kelly bar to blow the tailings out of the hole. This bit is used for small holes only.

After the VSM is in place, a slurry consisting of sand and water is poured into the foundation holes. The sand, which has been dried and heated beforehand, is delivered in insulated trucks to maintain its temperature. Once the slurry is in the hole, it freezes and hardens just as solidly as concrete.

Another product of TAPS research is a special saddle, or cradle, that carries the pipe on the cross member. Also kown as a slide-and-guide, the saddle compensates for lengthwise pipe movement caused by thermal expansion and contraction. A coated, stainless steel plate is welded directly onto the crossbeam, and a similar plate is also welded to the bottom of the saddle (fig. 5.35). The plates are coated with a low-friction polymer that allows them to move smoothly and freely as the pipe responds to temperature fluctuations.

Some VSMs used on the North Slope are not tall enough to permit caribou to pass beneath the pipeline. VSM height was therefore an important design consideration on the TAPS line, where it was feared that the pipeline would disrupt caribou migration. Arctic contractors regularly must construct special caribou crossings, which consist of gravel padding placed over the pipe.

EXPANSION LOOPS

The extent of thermal expansion and contraction can perhaps be fully appreciated only when it is realized that temperature differentials in the Arctic may range from −60°F to 80°F. Therefore, in conjunction with the slide-and-guide arrangement, expansion loops are intermittently built into the pipeline (fig. 5.36). An expansion loop is a configuration that permits the section to move without danger of overstressing the pipe.

CONSTRUCTION SEQUENCE

The sequence of Arctic construction is very much like that used in other areas. The differences, however, are that some of the usual steps are eliminated altogether, and that others are carried out in a way distinctive to above-ground Arctic lines.

The major restriction on Arctic pipe stringing is that it must be done during the winter or from gravel roads to help prevent damage to the tundra, as mentioned earlier. No bending is necessary on an Arctic spread, a fact attributable to both environmental and engineering considerations. All fittings and all bends for the expansion loops are fabricated in a heated shop and then transported to the right-of-way. All fabrication welds are numbered, and the final location of the piece and the acceptable tolerances are stringently controlled. Despite the apparent interchangeability of certain welds, such as the 90-degree pieces used in the expansion loops, each piece is tailored to an exacting degree for one location only.

Figure 5.36. Expansion loop

Stick electrode welding is the preferred technique on nearly all Arctic pipelines. Pipe ends must be preheated before the weld is made, and when done properly, the weld is unlikely to crack because of rapid cooling. Occasionally, an asbestos blanket is placed over the weld to slow its cooling, but this is not mandatory. X-ray inspection of all welds is mandatory, however, and standard techniques and equipment are used.

Arctic pipelines are insulated for two reasons. First, insulation prevents heat radiated by the warm oil within the line from thawing the permafrost beneath the pipe and allowing frost heaving. The second objective is the prevention of flow problems from excessive cooling of the oil. At extremely low temperatures, the viscosity of oil increases markedly, a fact well-known to anyone who has tried to start a car on a subfreezing winter morning.

Pipe joints are insulated at a mill before delivery to the right-of-way. Foam is pumped into an annulus formed between the outer pipe wall and a metal jacket that is fitted

Figure 5.37. Lowering up

around the circumference of the pipe. As the foam sets in the annulus, it hardens to a specific density. Since the bond formed between the pipe surface and the foam is strong and the pipe is not buried, a conventional pipe coating is not needed. Pipe ends left uncovered after welding are insulated in a similar fashion. Normally, a separate insulation crew places a mold over the joint. Foam is pumped through an opening in the top and is allowed to set.

Unlike conventional pipelining where the pipe is lowered into the ditch, the use of VSMs in Arctic work requires that the pipe be raised up onto the crossbeam. This lowering-up, as it is called, is done with either portable cranes or side booms (fig. 5.37).

In keeping with the regulations that stringently govern Arctic work, hydrostatic testing is required for all lines. If ambient temperatures are expected to remain above freezing for the duration of the test, water alone can be used to fill the line. For winter testing, however, a solution of 60 percent glycol and 40 percent water is reuired to prevent freezing in the pipe.

SPECIAL CONSIDERATIONS

Beyond the overriding emphasis placed on environmental protection, two other factors influence most Arctic pipeline work. The first is an increased incidence of equipment failure. As temperatures drop to around $-35°F$, construction equipment malfunctions multiply, especially in the case of hydraulic systems. A second consideration is the meticulous scheduling and planning required throughout the life of the project. The extensive documentation demanded by regulatory agencies is a major part of the long-term commitment to the pipeline's success. Equally important, though, is the need to keep a tight rein on expenses, which are enormous in the Arctic because of its remoteness and harsh climate.

Double Jointing

Double jointing is the process of welding two pipe joints together — usually on a double-joint rack — to form a single piece of pipe. The usual length of a double joint is 80 feet. Depending on the length of a single joint and the specifications of the job, however, double joints can be as long as 120 feet. Effective length is limited in large part by the distance from the yard to the right-of-way and the types of highways and access roads. Where public thoroughfares or roads are the only means of access, stringent regulations usually govern the transport of oversize loads.

Double jointing has several advantages over welding done in the field. The most obvious is that more time than usual can be devoted to quality-control measures. Defects can be detected and repaired immediately without the delays in other operations that would occur in the field. As a result, the total time and expense of welding is reduced, thus improving the pace of construction.

Welding is done in an area sufficiently large to accommodate the double-joint rack, auxiliary equipment, and the joints of pipe. A double-joint rack is a specialized piece of equipment consisting of two or more automatic welding stations connected by a combination of rails and rollers (fig. 5.38). The rack itself is portable and can be taken down and transported by truck to each job site. The initial expense of buying a double-joint rack, however, and the likelihood that its use may not prove feasible for every job, keep many contractors from purchasing one. Most opt to subcontract this operation to either a specialty company or another contractor.

Figure 5.38. Double-joint rack

In the interest of both safety and efficiency, the entire double-jointing process is highly automated. Pipe joints are brought in from outside storage by heavy equipment fitted with special pipe-handling attachments (fig. 5.39). Two joints at a time are placed on the rack and aligned. The line-up clamp holds the pipe joints in place as an automatic welder makes the first pass. When this is completed, the clamp retracts into a nest on the side of the rack well out of the way of the pipe as it rolls to the next station.

The sequence followed to complete the weld differs from that used in field welding; in fact, the order is almost entirely reversed. The first welding pass is made at roughly the midpoint on the bevel of the pipe end. The second pass is made in the same location as the cap bead—that is, on the outside of the bevel. Both of

Figure 5.39. Pipe being moved from yard to rack

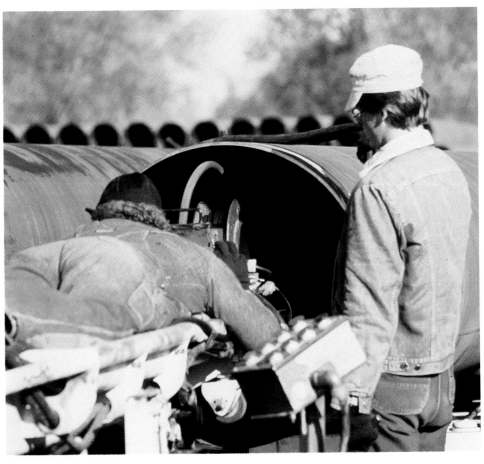

Figure 5.40. A crawler is used to position the welding machine in the inside of a double-joint weld.

these passes are welded automatically by a submerged-arc welding machine positioned above the pipe. The welding head is fixed at a precise angle and remains stationary as the joint is rotated.

The final pass is made on the inside of the pipe. To accomplish this, a welder—preferably one with a slim physique—rides into the pipe in a cradlelike crawler and positions the welding machine at the proper point (fig. 5.40). The welding machine remains stationary inside the pipe, and the pipe rotates around it. In addition to welding equipment, the crawler is outfitted with an exhaust fan to carry off welding fumes and draw in fresh air for the welder.

Beyond differences in the welding sequence, double jointing also uses amperages considerably higher than those available in the field. Pipeline welding uses any one of several types of electric-arc welding. Among other factors, the quality of the weld depends on the capacity of the power source producing the arc. On the spread this capacity is limited, since the benefits of greater power must be balanced against the need for welding units sufficiently portable to move easily and quickly along the right-of-way. The actual power delivered by a field unit varies, but it is usually in the range of 130 to 175 amperes. The welding heads used on double-joint racks, on the other hand, are able to maintain a power level of around 800 amperes. This level is comparable to those used in mill fabrication of pipe. The resulting welds are extremely fine grained and show greater strength and ductility.

Rehabilitation

As with most engineered or man-made structures, a certain amount of replacement or rehabilitation is necessary over a period of time. In the pipeline construction industry, much of the workload in any given year is devoted to this type of activity. Pipelines are replaced primarily due to three factors: population density near or over the line has increased substantially, corrosion has weakened the pipe, or the owner wishes to increase the capacity of the line.

The U. S. Department of Transportation's safety standards governing operating pressure requirements are based in part on population density. If an urban area expands so that the population near the pipeline right-of-way increases or if a building with a large number of occupants is constructed nearby, that section of the line must meet higher operating pressure standards. The owner can either reduce the operating pressure, thereby meeting the new safety standard although reducing throughput, or he can replace that section with pipe of heavier wall thickness and higher yield strength so that he can maintain his standard pressure. Corrosion too may mean that a portion of the line does not meet safety standards. In this case also, the owner may replace only the affected part of the line.

The construction sequence for rehabilitation work is essentially the same as for normal cross-country work, except that the existing pipe has to be removed. Removal may be one of the first steps if the owner shuts the line down so that the section can be taken out of service. On the other hand, if the owner wishes to reduce downtime, the contractor can lay the new section in a new ditch or on top of the ground before taking up the old section.

The existing line is usually stripped out with backhoes and uncovered. The pipe is taken up with side booms, cut up into 40-foot lengths, and hauled to a central stockpile area, where it is racked and prepared for resale. Preparation sometimes involves cleaning and straightening the old joints. With the old pipe out of the way, the ditch is cleaned and deepened, if necessary, and the new section is lowered in. This section is tested and then tied back into the existing adjacent pipeline.

Rehabilitation work often requires particularly close coordination among the various construction activities. The work is not strung out over many miles, as it is in normal cross-country work, but is condensed into a mile or so of highly congested right-of-way, and the normal spacing of operations is not possible. The superintendent who oversees rehabilitation procedures must be able to schedule the various crews and operations so that a minimum of downtime occurs.

Hydrostatic Testing

Most pipeline owners check the integrity of their lines annually. The most common form of inspection is hydrostatic testing. Usually done during periods of slack demand, this type of testing requires that the line be taken out of service and cut at specific locations. Manifolds are welded on, and the test sections are filled with clean water obtained from a nearby water source. The line is then tested in the same manner as for a new section. If no problems are encountered, then the line is dewatered, dried, and returned to service.

While the section is out of service for testing, the contractor, at the owner's request, performs any necessary remedial work. Remedial work often involves changing out valves, checking and repairing the coating, installing scraper traps and drips, replacing damaged sections of pipe, and other similar work. Routine testing, maintenance, and rehabilitation help to ensure that pipelines remain the safest form of transportation.

GLOSSARY

A

acetylene welding *n:* a method of joining steel components in which acetylene gas and oxygen are mixed in a torch to attain the high temperatures necessary for welding. An early type of welding, also called oxyacetylene welding. Its primary disadvantage was the seepage of molten weld material onto the interior surface of the pipe, often leading to corrosion problems.

aerial river crossing *n:* in pipeline construction, a river crossing technique in which the pipeline is either suspended by cables over the waterway or attached to the girders of a bridge designed to carry vehicular traffic.

anchor *n:* any device that secures or fastens equipment. In downhole equipment, the term often refers to the tail pipe. In offshore drilling, floating drilling vessels are often secured over drill sites by large metal anchors like those used on ships. For pipelines, a device that secures pipe in a ditch.

anchoring system *n:* in pipeline construction, a combination of anchors used to hold a lay barge on station and move it forward along the planned route. Lay-barge anchors may weigh in excess of 20 tons, and a dozen or more anchors may be needed.

auger *n:* a boring tool that consists of a shaft with spiral channels. An auger is used to bore the hole for a pipeline that must cross beneath a roadbed.

automatic welding *n:* a welding technique for joining pipe ends. Two general types of automatic welding used in pipeline construction are submerged-arc welding and automatic wire welding.

automatic wire welding *n:* an automatic welding process utilizing a continuous wire feed and a shielding gas. Automatic wire welding is similar to semiautomatic welding except that manual adjustment of the rate of wire feed and amount of shielding gas is unnecessary. See also *semiautomatic welding.*

B

backfilling *n:* the technique for covering a completed pipeline so that adequate fill material is provided underneath the pipe as well as above it. Backfilling prevents pipe damage due to loose rock, abrasion, shifting, and washouts.

backhoe *n:* an excavating machine fitted with a hinged arm to which is rigidly attached a bucket that is drawn toward the machine in operation. The backhoe is used for excavating and for clearing blasted rock out of the ditch during ditching for pipe laying.

barrel reamer *n:* in pipeline construction, a cylindrical device fitted on both ends with hollow cutting teeth, used in directionally drilled river crossings. Used during the pullback portion of a crossing effort, the barrel reamer opens the hole and keeps the pull on course.

Big Inch *n:* the first cross-country pipeline with a 24-inch diameter. The 1,340-mile Big Inch was begun in 1942 with government financing as a part of an emergency construction program to meet the demand for petroleum products during World War II (War Emergency Pipelines).

big-inch pipe *n:* thin-walled pipe of high tensile strength with diameters of 20 inches or more.

blasting mats *n pl:* coverings used to contain the onslaught of flying debris and rock caused by the use of explosives during pipeline ditching.

bore *v:* to penetrate or pierce with a rotary tool. Compare *tunnel.*

bottom-pull method *n:* an offshore pipeline construction technique in which the pipe string remains below the surface while it is towed to its final location.

bury barge *n:* in pipeline construction, a barge used for trenching underwater lines. Like a lay barge, a bury barge depends on an anchoring system for propulsion.

C

canal-lay construction *n:* a pipeline construction technique used in swamps and marshes. The first of several barges clears the right-of-way and digs a trench large enough to float itself and the following barges.

cap bead *n:* the final welding pass made to complete the uniting of two joints of pipe.

carrier pipe *n:* term used to refer to a pipeline when other pipe, called casing, is used with it in crossing under roadbeds and railroad right-of-ways. See *casing.*

casing *n:* large pipe in which a carrier pipeline is contained. Casing is used when a pipeline passes under railroad right-of-ways and some roads to shield the pipeline from the unusually high load stresses of a particular location. State and local regulations identify specific locations where casing is mandatory.

cathodic protection *n:* a means of preventing the destructive electrochemical process of corrosion of a metal object by using it as the cathode of a cell with a sacrificial anode. Current at least equal to that caused by the corrosive action is directed toward the object, offsetting its electrical potential.

cleaning and priming machine *n:* a self-propelled machine that removes from pipe surface any loose material with a rotating set of brushes and buffers and applies a thin coat of primer to prepare the pipe for coating.

clear *v:* to remove brush, trees, rocks, and other obstructions from an area.

coating flaw *n:* a gap or flaw in pipe coating. Also called holiday. Coating flaws, which must be repaired to prevent corrosion problems, are detected through mechanical or visual inspections of the line.

coating machine *n:* a machine that applies an even layer of coating material to pipe surface. Most coating machines also apply outer wrapping on the pipe.

cold-work *v:* to work metal without the use of heat. Compare *fire bending.*

conventional river crossing *n:* a waterway crossing in which pipeline construction techniques similar to those on land are used. In a conventional crossing, the route is graded and ditched; the pipe is welded, tested, and pulled across the stream; and then the pipe is tied in on each side to the cross-country line.

corrosion *n:* any of a variety of complex chemical or electrochemical processes by which metal is destroyed through reaction with its environment. For example, rust is corrosion.

cover depth *n:* the measurement from the top of a pipe to ground level along a right-of-way. Ditch depth and cover requirements are regulated by the U.S. Department of Transportation.

crawler *n:* a self-propelled X-ray machine that rides inside pipe to examine welds for possible defects.

cut and fill *v:* to cut down high ground or fill in low ground to achieve a uniform grade for a pipeline.

cutterhead *n:* in pipeline construction, the lead component in a directional drilling assembly. A circular steel band ringed with conical cutting teeth, the cutterhead does the actual boring of the hole for the pipeline under the waterway being crossed. Also called fly cutter.

D

deadman *n:* an anchoring point against which the winch on a boring machine for pipelining can pull.

deflect-to-connect connection *n:* an underwater pipe-joining technique in which the pipe is pulled to a target area in line with the platform but to one side of it. The connection is made by winding or otherwise deflecting the pipe laterally until it mates with the riser connection. Compare *direct pull-in connection.*

destructive testing *n:* a procedure in which a weld is torn apart so that its structure can be examined. Destructive testing is used primarily during the qualification procedures required of all welders who work on pipelines.

directional drilling *n:* a technique of river crossing in pipeline construction in which the pipe is buried under the riverbed at depths much greater than those of conventional crossings. With this technique, a hole in the form of an inverted arc is drilled beneath the river, and the actual made-up pipeline is pulled through it.

direct pull-in connection *n:* an underwater pipe-joining technique in which the pipe with its connector sled is steered straight onto the matching receiver at the base of the platform. Compare *deflect-to-connect connection.*

disbonding *n:* a common coating failure in which the coating separates from the pipe.

ditch *v:* to excavate a trench in which to lay pipe or cable. Ditching equipment and methods for pipe laying vary according to terrain and weather encountered.

ditch breaker *n:* a device that divides a ditch into sections to form internal barriers to water movement. Ditch breakers are used for pipelines in areas where washouts are a threat.

dope pot *n:* a portable container used to melt coal tar enamel and maintain it at the temperature required by a pipe-coating operation.

double jointing *n:* the process of welding two pipe joints together—usually on a double-joint rack—to form a single piece of pipe. The usual length of a double joint for pipeline construction is 80 feet.

dummy pipe *n:* the pipe used to slick-bore a road. The dummy pipe is bored through first to create the hole, and the carrier pipe is welded to it. The dummy is pulled through, thereby positioning the carrier pipe.

dynamic tensioning *n:* a sophisticated monitoring system for laying pipe, used to control pipe release off a stinger. This system compensates immediately for horizontal and vertical wave motion by either paying out the pipe string or hauling it back. Dynamic tensioning maintains a constant level of tension on the pipe and prevents excessive stress on the line.

E

easement *n:* a right that one individual or company has on the surface of another's land. In the petroleum industry, it usually refers to the permission given by a landowner for a pipeline or access road to be laid across his property.

enamel coating *n:* a collective term for a variety of petroleum-based derivatives such as asphalts, coal tars, grease and wax, mastics, and asphalt mastics that are used to coat pipe.

expansion loop *n:* a full loop built into a pipeline to allow for expansion and contraction of the line.

external line-up clamp *n:* an alignment clamp used on the outside of pipe. External line-up clamps are usually used on pipe with a diameter of 8 inches or less. Compare *internal line-up clamp.*

F

fabrication *n:* a collective term for the specialized connections and fittings on a pipeline. Fabrication assemblies control product flow, direct products to the proper location, aid in product separation, and facilitate maintenance operations.

fabrication crew *n:* pipeline construction workers responsible for welding fabrication assemblies into the line. The fabrication crew works independently from the rest of the spread.

field administration *n:* in pipeline construction, the superintendent and crew foremen.

field bevel *n:* a rebeveling of pipe ends in the field, usually required because of damage sustained by the pipe during transport or because a defective weld must be cut out.

field office manager *n:* the individual responsible for the contractor's financial affairs on a pipeline spread. The field office manager oversees billing arrangements, payroll, and other money-related matters.

fencing crew *n:* pipeline construction workers responsible for constructing temporary gates at points where a right-of-way crosses fence lines.

field support personnel *n:* in pipeline construction, the mechanics, parts and warehouse workers, truck drivers, and others who service the machinery that actually does pipe laying.

fire bending *n:* one of the earliest methods for bending pipe. The joint was first placed over a small bonfire, and when the heat had rendered it sufficiently malleable, it was placed against a tree, and pressure was applied until the desired bend was achieved. Fire bends significantly weakened the pipe. A cold-work process is less damaging.

firing line *n:* in pipeline construction, the welding crew that takes over after the root pass and the hot pass have been made. The firing line is responsible for the filler pass and the cap bead, which complete the joint.

flash welding *n:* in pipeline construction, a welding technique in which low voltage is applied to each pipe joint while the ends are in light contact. This contact produces a rapid arcing, called flashing. After the pipe ends have been adequately heated, the current is abruptly increased, and the pipe joints are brought together rapidly and forcefully. The current is then reduced, excess flash material in the pipe is cleared, and the weld is completed.

flow-control connection *n:* a device that controls product flow and directs it to the proper location. Mainline valves and side taps are examples of flow-control connections.

flume pipe *n:* large pipe used in creek and stream ditching in pipeline construction to allow the water to flow normally and provide a passage for equipment over the water.

fly cutter *n:* See *cutterhead.*

freeze pipe *n:* a device fitted on the vertical support members of the TAPS pipeline to circulate a refrigerant continuously between the subsoil and the top of the pipe. The refrigerant keeps the ground beneath the pipeline frozen to prevent frost heaving.

frost heaving *n:* movement of the soil resulting from alternate thawing and freezing. Frost heaving generates stress on vertical support members of pipelines in the Arctic and, by extension, on the pipe itself.

fusion-bonded epoxy coating *n:* a powdered resin coating that forms a skin over pipe when applied to its heated steel surface. Fusion-bonded epoxy coatings are usually applied at the mill.

G

grading *n:* the process of providing a smooth and even work area to facilitate the movement of equipment onto and along a right-of-way. Grading entails leveling, cutting, and filling.

grinding and buffing *n:* in pipeline construction, the process of cleaning pipe ends of dirt, rust, mill scale, or solvent to prepare them for welding. Grinding and buffing tasks are accomplished with power hand tools such as wire brushes and buffers.

guidance system *n:* in pipeline construction, the means by which a river crossing operation stays on course. Frequently computerized, guidance systems may be based on information gathered by pendulants, which determine inclination; gyroprobes, which are sensitive to drift and bearing; and sonar. Lasers are also used to guide crossings.

H

holiday *n:* a gap or void in coating on a pipeline or in paint on a metal surface.

holiday detector *n:* an electrical device used to locate a weak place, or holiday, in coatings on pipelines and equipment. Also called jeep.

hot pass *n:* the second pass made on a weld. The hot pass follows the root, or stringer, bead and precedes the filler pass and cap.

hot tie-in *n:* a weld made on a pipeline already in service. In the hot tie-in procedure, the gas from the line is purposely ignited at the point where the welding is to be done. Igniting the gas eliminates the chance that a spark could cause an explosion of gas and air.

hydrostatic testing *n:* the most common final quality-control check of the structural soundness of a pipeline. In this test, the line is filled with water and then pressured to a designated point. This pressure is maintained for a specific period of time, and any ruptures or leaks revealed by the test are repaired. The test is repeated until no problems are noted.

I

internal line-up clamp *n:* an alignment clamp used on the inside of pipe. The internal line-up clamp uses a number of small expandable blocks, or shoes, to grip the inside surfaces of both pipe joints and hold them in place. The clamp can also act as a swab to clean the inside of the pipe. Compare *external line-up clamp.*

J

jeep *n:* also called holiday detector. See *holiday detector*.

jet sled *n:* in pipeline construction, a pipe-straddling device, fitted with nozzles on either side, that is towed by a bury barge. As water is pumped at high pressure through the nozzles, spoil from beneath the pipe is removed and pumped to one side of the trench. The line then sags naturally into position in the trench.

joint *n:* in pipelining, a single length (usually 40 feet) of pipe.

J curve *n:* the configuration of pipe when it enters the water from an inclined ramp on the stern of a lay barge instead of from a stinger. The J curve eliminates overbend, which can stress the pipe.

J-tube method *n:* a method for joining a pipeline to a subsea riser on a platform. In the J-tube method, the pipe is lifted off the ocean floor when it reaches the platform and is then fed up to the surface through a guide tube. During this process, the pipe assumes a J configuration. Compare *reverse J-tube method*.

L

lay barge *n:* a barge used in the construction and placement of underwater pipelines. Joints of pipe are welded together and then lowered off the stern of the barge as it moves ahead.

lay-barge construction *n:* a pipe-laying technique used in swamps and marshes in which the forward motion of the barge sends the pipe down a ramp and into the water. Also called *marine lay*.

line travel applied coating *n:* in pipeline construction, the coating applied to pipe over the ditch. Coal tar enamels are a particularly effective type of line travel applied coating.

Little Big Inch *n:* a 20-inch products line constructed during the same period as Big Inch as part of the World War II effort. See *Big Inch*.

looping *n:* the technique of laying an additional pipeline alongside an existing one when additional capacity is desired.

lowering-in *n:* the process of laying pipe in a ditch in pipeline construction. Pipe can be lowered into the ditch as part of the coating operation or lowered separately by a lowering-in crew.

lowering-up *n:* in pipeline construction, the process of raising pipe and placing it on vertical support members in parts of the world where frozen earth prevents normal burial of the line. Lowering-up is the counterpart of lowering-in in more temperate climates.

M

manual welding *n:* a welding process in which an electric arc melts and fuses the pipe ends with the metal of an electrode held by the welder. Also called stick welding.

marine lay *n:* See *lay-barge construction*.

mill-coated pipe *n:* pipe coated at the mill as opposed to pipe coated over the ditch in pipeline construction.

N

nondestructive testing *n:* in pipeline construction, testing designed to evaluate the quality of both production and field welds without altering their basic properties or affecting their future usefulness. The most common nondestructive testing is radiographic, or X-ray, testing. Compare *destructive testing*.

O

offshore pipeline construction *n:* pipeline construction in water depths of 100 feet or more.

open-cut crossing *n:* a road crossing in which the pipeline ditch cuts through the road instead of being bored under it. An open-cut crossing is generally used in sparsely populated areas where the right-of-way crosses little-used dirt or gravel roads. Open-cut crossings are more convenient than bored crossings and also hold down costs.

oxyacetylene welding *n:* See *acetylene welding*.

P

padding *n:* screened or sifted dirt, clean gravel, or foam placed in a ditch to protect pipe from damage caused by rocky or rough soils.

Pearson holiday detector *n:* a holiday detector that checks for coating defects as well as any metal debris near a buried pipeline.

pig *n:* in hydrostatic testing of a pipeline, a scraper used inside the line to push air out ahead of the test water and to push water out after the test.

pig run *n:* the trip of a pig through a pipeline. See *pig*.

pilot hole *n:* in pipeline construction, the hole drilled as the first step of a directionally drilled river crossing. The pilot hole establishes a pathway for the pipeline.

pilot string *n:* joints of small-diameter pipe attached to the drill assembly used to bore a pilot hole in laying pipe. After the route has been established, the pilot string is replaced by the work string.

pipe bending *n:* the process of bending joints so that a pipeline will conform to the topography of a right-of-way. Pipe bends are made by the cold-work process.

pipe-bending machine *n:* in pipeline construction, a track-mounted, hydraulic machine that bends a joint to the precise angle specified by the bending engineer. The bend is made by a set of clamps that grip the outside surface of the pipe and prevent slippage while a winch cable hooked to the free end of the pipe maintains upward pull and guides the pipe through the machine.

pipe coating *n:* a special material that coats pipe for pipelines and prevents water from coming into contact with the steel of the pipe. The most widely used types of pipe coatings are bituminous enamels, epoxy resins, and tapes.

pipe gang *n:* in pipeline construction, the workers responsible for positioning the pipe, aligning it, and making the initial welds. The pipe gang sets the pace that determines the progress of the rest of the spread.

pipeline *n:* a system of connected lengths of pipe, usually buried in the earth or laid on the seafloor, that is used for transporting petroleum and natural gas. A pipeline serves as both a conveyor and a temporary container.

pipeline testing *n:* the process of proving the structural soundness of an installed pipeline and its capability to fulfill safely the function for which it was designed. The most common testing method is hydrostatic testing.

pipe tensioner *n:* a braking device used on a lay barge to control the descent rate of the pipe. Tensioners also support the entire submerged weight of the pipe as it approaches the bottom.

pipe wrapping *n:* material applied on top of pipeline coating to protect the coating from damage. Materials used for wrapping include felt, fiberglass, fiberglass-reinforced felt, and kraft paper.

preheating *n:* in pipeline construction, the process of heating pipe ends before welding. Preheating is usually necessary in areas where ambient temperatures are below 40°F or where there is overnight condensation of moisture on the pipe. Wagon-wheel heaters are used for preheating.

pup joint *n:* a length of drill or line pipe, tubing, or casing shorter than 30 feet.

push-in construction *n:* a pipe-laying technique used in swamps and marshes in which the pipe itself is moved from a stationary point out into the ditch. Compare *lay-barge construction.*

R

radiographic testing *n:*1 photographic record of corrosion damage obtained by transmitting X rays or radioactive isotopes into production structures. It may also be used to produce a shadowgraph of a pipeline weld and reveal any flaws. Also called X-ray testing.

reel barge *n:* a lay barge specially outfitted to lay pipe from an immense reel on deck.

reel method *n:* an offshore pipeline construction technique in which the welded, coated, and tested pipe is coiled onto a reel and transferred to a reel barge where it is paid out at a steady rate onto the ocean floor.

remote connection *n:* an offshore pipe-joining technique in which the connection process is directed from somewhere other than at the immediate site, such as from a control panel on the platform deck. Two types of remote connection are the direct pull-in and the deflect-to-connect techniques.

reverse J-tube method *n:* a method of joining a pipeline to a subsea riser on an offshore platform. In this method the pipe is welded together on the platform itself and then fed down through a guide tube to the seafloor. Compare *J-tube method.*

right-of-way *n:* the legal right of passage over public land and privately owned property; also the way or area over which the right exists. The width of a right-of-way varies according to contract specifications and individual easements, but it is generally between 50 and 100 feet.

right-of-way restoration *n:* in pipeline construction, the process of returning a right-of-way to its original condition or better after the pipeline has been completed. Right-of-way restoration depends on legal stipulation in the contract with the pipeline owner and agreements made with individual landowners.

ripper *n:* a claw-shaped, plowlike attachment used on a bulldozer to loosen rock and locate solid formations that may require explosives in clearing a right-of-way for pipeline construction.

riprap *v:* to space logs and timbers evenly along the length of a pipeline right-of-way to stabilize soil in swamps or wet areas.

river crossing *n:* a type of special pipeline construction used when a pipeline must cross a river or stream. Types of river crossings include aerial crossings, conventional crossings, and directionally drilled crossings.

road crossing *n:* laying of a pipeline under a roadbed or through a road.

rock ditching *n:* excavating a trench in rock or rocky soil.

rock drill *n:* a drill generally powered by compressed air and used to drill holes for explosives. Rock drills may be required in pipeline right-of-way grading and in ditching.

root bead *n:* the initial welding pass made in uniting two pipe joints. Also called stringer bead.

ROW *abbr:* right-of-way.

S

sag bend *n:* a temporarily unsupported span of pipe between the stinger and the seabed in marine pipe laying.

saturation diving *n:* diving in which a diver's tissues are saturated with an inert gas to a point where no more of the gas can be absorbed by his body. Consequently, once a diver is saturated, decompression time remains the same whether he stays at the saturated depth for 24 hours or for several days.

scraper *n:* any device that is used to remove deposits (such as scale or paraffin) from tubing, casing, rods, flow lines, or pipelines.

S curve *n:* the configuration of pipe when it enters the water from a stinger of a lay barge in pipeline construction. The overbend is closest to the barge, and the sag bend is on the seafloor. Compare *J curve*. See *sag bend*.

semiautomatic welding *n:* a welding technique in which the arc is maintained in a continuous stream of gas between an electrode and the pipe being welded. The semiautomatic welding apparatus consists of a spool of coiled wire that provides filler metal for the weld, a pair of driving rollers that guide the wire to the weld, a welding gun, and a supply of shielding gas. Each type of semiautomatic welding is usually identified by the type of shielding gas used; in CO_2 wire welding, for example, carbon dioxide is the shielding gas.

semisubmersible lay vessel *n:* a type of pipe-laying vessel with a submerged pontoon hull and an elevated work area. Semisubmersible lay vessels remain relatively steady in high seas and are used in areas where conditions are expected to be continually rough.

shielded-arc welding *n:* a welding technique in which the rod coating involves an inert gas shield that protects the weld from the rapid oxidation caused by contact with oxygen in the atmosphere. Shielded-arc welds are fine grained, are free of oxides and nitrides, and have great ductility and toughness.

shielding gas *n:* an inert gas that is used as a shielding medium in pipe welding. Its primary purpose is to prevent oxidation of the weld at the point of contact with the pipe metal by excluding oxygen in the air from the area around the molten metal.

shoofly *n:* a special access road constructed to link a right-of-way with existing roads. Shooflies are necessary only in remote areas.

shooting rock *n:* the process of using explosives to clear rock from a pipeline right-of-way or from the ditch line. Also called blasting.

slack looping *n:* the process of laying pipe alternately on opposite sides of a ditch to counter the effects of pipe contraction and expansion caused by extreme variations in daily temperatures.

slick boring *n:* in pipeline construction, a boring technique sometimes used for road crossings in which a large amount of liquid is pumped into the hole outside of the pipe to reduce friction.

slide-and-guide *n:* a special saddle, or cradle, that holds pipe on a vertical support member in the lowering-up type of pipe laying. The slide-and-guide allows longitudinal movement of the pipe caused by thermal expansion and contraction.

slurry *n:* a mixture in which solids are suspended in a liquid.

spacer *n:* member of a pipeline construction gang who is responsible for assuring that the exact distance between the beveled pipe ends for a welding process to be used on the joint is maintained. Spacers strike a wedge into the interface between the pipe bevels and then maneuver them to an exact, uniform distance around the entire circumference.

spoil *n:* excavated dirt.

spread *n:* the necessary equipment and crew needed to build a pipeline. Modern spreads, which are like moving assembly lines, can consist of one hundred pieces of equipment and over five hundred workers.

spreader bar *n:* in pipeline construction, a rod positioned between two lifting lines so that the weight of a joint being lifted off the trailer is evenly distributed and that the pipe does not buckle in the center.

spread superintendent *n:* in pipeline construction, the individual with responsibility for running the spread. The spread superintendent represents the contractor's interests in the field.

stick welding *n:* See *manual welding*.

stinger *n:* a device for guiding pipe and lowering it to the water bottom as it is being laid down by a lay barge. The stinger is hinged to permit adjustments in the angle of pipe launch.

stovepipe assembly *n:* in laying pipe, an assembly on lay barges in which pipe joints are assembled in a continuous string. Each joint passes through individual work stations spaced along a gently sloping production ramp.

stringer bead *n:* See *root bead*.

stringing *n:* in pipeline construction, the process of delivering and distributing line pipe where and when it is needed on the right-of-way. Stringing also includes the delivery of joints of special wall thickness and pipe grade to specific locations such as road crossings where heavy wall thickness may be specified by the contract or by regulations. Pipe is strung so that the movement of livestock and vehicles is not impeded.

submerged-arc welding *n:* an automatic welding process utilizing a continuous wire feed and a shielding medium of fusible granular flux. Submerged-arc welding offers high deposition rates and weld passes of substantial thickness. It is used primarily in double-joint racks.

submersible *n:* a two-man submarine used for inspection and testing of offshore pipelines.

subsea riser *n:* a vertical section of pipe that connects pipeline on the sea bottom to a production platform on the surface. The riser is an integral part of the pipeline and is clamped directly to a leg or brace on the platform.

suction dredge *n:* in pipe laying, a type of trenching machine used on river crossings when the channel cannot be diverted or when the volume of material to be removed is large. A suction pump rapidly forces large amounts of soil into a discharge pipe for deposit on the adjacent bank. A cutterhead can also be used on a suction dredge.

T

tape coating *n:* in pipeline construction, a protective coating of polyethylene, polyvinyl, coal-tar-base, or butyl-mastic tape that is wrapped around pipe to prevent corrosion. Tape coatings are applied on the line, and any defects can be repaired relatively quickly and simply.

taping machine *n:* in pipeline construction, a machine that moves along the pipe, wrapping joints with tape in overlapping segments.

TAPS *abbr:* Trans-Alaska Pipeline System.

tie-in *n:* a collective term for the construction tasks bypassed by regular crews on pipeline construction. Tie-in includes welding road and river crossings, valves, portions of the pipeline left disconnected for hydrostatic testing, and other fabrication assemblies, as well as taping and coating the welds.

tie-in gang *n:* workers responsible for tie-in tasks in the construction of a pipeline.

topsoiling *n:* the technique of placing topsoil in a spoil bank separate from the rest of the excavated materials from pipeline construction so that it can be replaced in its original strata during backfilling operations.

tow cat *n:* a tractor used to tow equipment up steep grades.

Trans-Alaska Pipeline System *n:* the largest long-distance, big-inch pipeline in the noncommunist world. Stretching approximately 800 miles from Prudhoe Bay to Valdez, Alaska, the 48-inch pipeline also known as TAPS, was completed in 1977 at a cost of $8 billion.

tundra *n:* the treeless plain characteristic of much of the Arctic. Although the tundra remains frozen for much of the year, it is highly sensitive to environmental disturbance, and special pipeline construction techniques are required there.

tunnel *v:* in pipeline construction, to penetrate or pierce manually. Compare *bore*.

U

uncased crossing *n:* in pipeline construction, a road crossing bored without casing. In an uncased crossing, the carrier pipe itself is pushed under the roadbed by the boring machine. To reduce potential damage to the coating, slick boring is frequently used. See *slick boring*.

V

vent *n:* a device installed on one end of that portion of a pipeline that crosses under a road. The vent marks the boundary of the highway right-of-way and provides an exit for any fluids should the pipeline develop a leak. It also aids in locating line breaks.

vertical support member *n:* H-shaped device that supports a pipeline above the ground. Vertical support members are used in parts of the world where frozen earth prevents normal burial of the line. See *lowering-up*.

VSM *abbr:* vertical support member.

W

wagon-wheel heater *n:* in pipeline construction, a multiheaded circular propane torch that is manually rotated inside the end of a pipe for preheating.

War Emergency Pipelines *n:* a government-financed, nonprofit corporation of eleven oil and pipeline companies established during World War II to build desperately needed pipelines such as Big Inch and Little Big Inch.

well point *n:* in pipeline construction, a device installed in waterlogged soil to dry out areas along the ditch line. Functioning as a submersible pump, a well point is a hollow steel rod approximately 18 feet long driven into the ground. Its free end is connected by hose or tubing to a well-point pump that moves groundwater away from the ditch. Hundreds of well points may be required to stabilize an area.

WEP *abbr:* War Emergency Pipelines.

wet boring *n:* in pipeline construction, a boring technique similar to slick boring that is used for small-diameter pipe such as natural gas distribution lines. Water is the lubricant in wet boring. Compare *slick boring*.

wheel ditcher *n:* a ditching machine for pipeline construction that has a large, rapidly rotating set of toothed scoops that lift dirt out of the ditch and feed it onto a conveyor mounted on the side of the machine. The wheel ditcher is used almost exclusively for ditching operations in stable soil.

work string *n:* in pipeline construction, the string of washpipe that replaces the pilot string in a directionally drilled river crossing. The work string remains in place under the river until the actual pipeline is made up and is ready to be pulled back across the river.

X

X-ray testing *n:* See *radiographic testing*.